The Logical Status of Diagrams

The Logical Status of Diagrams

SUN-JOO SHIN
University of Notre Dame

CAMBRIDGE UNIVERSITY PRESS
Cambridge, New York, Melbourne, Madrid, Cape Town, Singapore, São Paulo, Delhi

Cambridge University Press
The Edinburgh Building, Cambridge CB2 8RU, UK

Published in the United States of America by Cambridge University Press, New York

www.cambridge.org
Information on this title: www.cambridge.org/9780521102773

First published 1994
This digitally printed version 2009

A catalogue record for this publication is available from the British Library

Library of Congress Cataloguing in Publication data

Shin, Sun-Joo
The logical status of diagrams / Sun-Joo Shin.
p. cm.
Includes bibliographical references.
ISBN 0-521-46157-X
1. Venn diagrams. 2. Logic, Symbolic and mathematical.
I. Title.
QA248.S423 1994
511.3′3–dc20 94-8822 CIP

ISBN 978-0-521-46157-3 hardback
ISBN 978-0-521-10277-3 paperback

For My Parents

Contents

Acknowledgments

I have received so much help from many people throughout various stages of this work. Among them, I particularly express great thanks for John Perry's endless support and excellent suggestions for this work. Jon Barwise's patience and support made it possible for me to formulate the main idea of this project. His help has been crucial at every stage. Above all, I want to say here that I had the best advisor, John Etchemendy, while I was a graduate student at Stanford. He has been more helpful in every way than anyone could imagine an advisor can be, even after graduation. Without his support, I could never have survived my frustrations. In addition, I would like to thank Hans Kamp, Gordon Plotkin, and the anonymous reviewers of Cambridge University Press for their encouragement and suggestions. I also feel grateful to Michael Bratman who gave me a chance to visit Stanford while working on this project. I received many kinds of help from those working in the Department of Philosophy and the Center for the Study of Language and Information at Stanford, particularly Ingrid Deiwiks, Dikran Karagueuzian, Michele King, Betsy Macken, Emma Pease, Nancy Steege, and Zita Zukowsky. I also give thanks to the Philosophy Department of the University of Notre Dame for their understanding and support for this work. Lauren Cowles, Louise Calabro Gruendel, Pauline Ireland, Gregory Schreiber, and Alison Woollatt at Cambridge University Press have been a great help.

I give most special thanks to my mother who has always prayed for me in Korea and to my husband, Henry Smith, for his great understanding, patience, support, and love. His discussion of my work has always been one of the most fruitful for making this work possible.

1
Introduction

Diagrams have been widely used in reasoning, for example, in solving problems in physics, mathematics, and logic. Mathematicians, psychologists, philosophers, and logicians have been aware of the value of diagrams and, moreover, there has been an increase in the research on visual representation. Many interesting and important issues have been discussed: the distinction, if any, between linguistic symbols and diagrams, the advantages of diagrams over linguistic symbols, the importance of imagery to human reasoning, diagrammatic knowledge representation (especially in artificial intelligence systems), and so on.

The work presented in this book was mainly motivated by the fact that we use diagrams in our reasoning. Despite the great interest shown in diagrams, nevertheless a negative attitude toward diagrams has been prevalent among logicians and mathematicians. They consider any nonlinguistic form of representation to be a heuristic tool only. No diagram or collection of diagrams is considered a valid proof at all. It is more interesting to note that nobody has shown any legitimate justification for this attitude toward diagrams. Let me call this traditional attitude, that is, that diagrams can be only heuristic tools but not valid proofs, the general prejudice against diagrams. This prejudice has been unquestioned even when proponents of diagrams have worked on the applications of diagrams in many areas and argued for the advantages of diagrams over linguistic symbols. This is why it is quite worthwhile to question the legitimacy of this prejudice, that is, whether this prejudice is well grounded or not.

This prejudice has been explicitly expressed in mathematicians' and philosophers' writings. When Hilbert says the following about his own diagrammatic demonstration of Jordan's theorem, he obviously shares the common mistrust of diagrams as proof:

I have given a simplified proof of part (a) of Jordan's theorem. Of course, my proof is completely arithmetizable (otherwise it would be considered non-existent); ...[1]

If Hilbert had arithmetized a proof first, he would not have said, "My proof is completely diagrammatizable (otherwise it would be considered non-existent)." Eisenberg and Dreyfus recollect a seminar on visualization in presenting mathematics (held at Tel Aviv University in 1987) in the following way:

But it soon became apparent that these diagrams were not considered proofs at all – but something much less in status – mathematical mnemonics. ...The consensus of opinion was that although the diagrams could be used to generate a proof they certainly were not proofs in and of themselves.[2]

In the same paper, they confirm this problem and identify a clear prejudice against visualization in the tradition of mathematics:

...and it is our contention that students (and oftentimes their teachers too) cannot do many of the problems in this list because they have not learned to exploit the visual representations associated with the concepts. ...The point is that the visual characteristics of the problem were not even considered. And that seems to be exactly the problem; visual aspects of a concept are rated secondary to the concept itself.[3]

These authors find in the following an admission that the general belief is that a visual proof is not a real proof:

The belief that a visual proof is not a mathematical proof should be eliminated.[4]

The following passage from Tennant clearly shows a prejudice for linguistic representation and, moreover, the assumption about proofs that they do not include a diagrammatic method at all:

[The diagram] is only an heuristic to prompt certain trains of inference; ...it is dispensable as a proof-theoretic device; indeed, ...it has no proper place in the proof as such. For the proof is a syntactic object consisting only of sentences arranged in a finite and inspectable array.[5]

The previous quotations are examples of the general prejudice against visualization correctly diagnosed by Barwise and Etchemendy:

[1]Quoted by Hadamard [12], p. 103.

[2]Eisenberg and Dreyfus [8], p. 10.

[3]Eisenberg and Dreyfus [8], pp. 1–2.

[4]S. Vinner, "The Avoidance of Visual Considerations in Calculus Students," *Focus: On Learning Problems in Mathematics* 11, 2 (1989) p. 156, quoted in Eisenberg and Dreyfus [8].

[5]Neil Tennant [26], p. 304, quoted by Barwise and Etchemendy [4], p. 80.

But despite the obvious importance of visual images in such human cognitive processes, visual representation remains a second-class citizen in both the the theory and practice of mathematics. In particular, we are all taught to look askance at proofs that make crucial use of diagrams, graphs, or other non-linguistic forms of representation, and we pass on this disdain to our students.[6]

During the past hundred years, logicians have developed an arsenal of techniques for studying valid inference in the case where all the information is expressed in sentences of a formal language. ...These familiar techniques [various deductive systems and semantic techniques] have been tailor-made for the homogeneous, linguistic case.[7]

Where does this prejudice come from? I find there are two main sources that are responsible for this prejudiced opinion about diagrams. One is the limitation of diagrams in representing knowledge. The other, which has been taken more seriously than the first one, is the possibility of misuse of diagrams. I will now address the question of whether these two factors are sufficient enough to justify the general prejudice against diagrams with which we are concerned.

It is true that diagrams have their own limits in representing information. We cannot represent everything we want by means of diagrams. In this sense, I agree with any regretful remark about the limited expressive power of diagrammatic representation. However, one cannot move from this remark to the extreme conclusion that no diagram can be used as a valid proof, without assuming that to be valid a representation should not be limited in its power of representing information. But we know that this assumption cannot be sustained. If anyone believes this assumption, he should not count a proof in first-order logic as a valid proof either. As all of us are aware, there are many things that cannot be expressed in first-order logic. Nevertheless, none of us would stop using first-order logic for a valid proof when first-order logic can be applied. Accordingly, it is not legitimate to prevent diagrams from being used as valid proofs just because there are many things that diagrams cannot express. This discussion shows that the first charge against diagrams – diagrams cannot represent everything – is true in itself, but it cannot support the general prejudice against diagrams.

The long history of mistrust of diagrams stems mainly from the second complaint against diagrams that the use of diagrams tends to mislead us in reasoning. Especially in geometry, there have been constant warnings and great concern about accidental properties of geometrical figures.

[6]Barwise and Etchemendy [2], p. 9.
[7]Barwise and Etchemendy [3], p. 33.

For example, if you wanted to prove some property about a triangle in general, but the triangle you drew on the paper happened to be isosceles, then you might easily commit a fallacy in your proof by using some accidental properties of this triangle. Or, sometimes a diagram can mislead us into denying a true proposition, as in the following example.[8]

If, in the figure, $AB \cong AD$, and $BC \cong CD$, then $\angle AEB$ is a right angle.

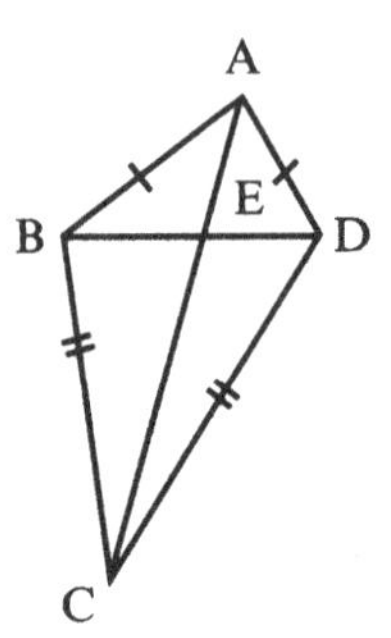

In the diagram, the $\angle AEB$ does not look like a right angle. Suppose someone thinks that this diagram itself is a counterexample to this proposition and concludes that this proposition is wrong. A fallacy results from misplaced reliance on the diagram.

As these examples show, the warnings against misapplication of diagrams are extremely important. However, the possibility of misusing accidental properties does not justify excluding diagrams from valid proofs altogether. In the first example, we would simply point out to the reasoner that if he aims to prove some property of a triangle in general he should not use a special property of a special kind of a triangle, in that case, an isosceles triangle. What this mistake tells us is *not* that we should not have used a diagram, *but* that we should not have used a special property of an isosceles triangle or at most that we should not have drawn an isosceles triangle. Likewise, in the second example, we would point out that the appearance of the size of an angle or of the length of a line should not be trusted. That is, these appearances are not among the properties we can use for a valid proof.

Diagrams, however, are not alone among representation systems in having such accidental properties that should not be used in a proof. In fact, in any kind of representation system, there are accidental properties. If someone takes a difference in sizes of letters in symbolic logic to be a

[8]Brumfiel, Eicholz, and Shanks [6], p. 105.

significant fact, then he would be led to commit fallacies. This is a main reason why we need to know the syntax and the semantics of a system in order to understand and use it. The traditional complaint against diagrams, however, goes further: Since diagrams are misused, we should not use them in a valid proof at all. The traditional moral would seem to be that any system which can possibly be misused should not be used for any valid proof. Barwise and Etchemendy point out that this assumption would lead to disaster:

> If we threw out every form of reasoning that could be misapplied by the careless, we would have very little left. Mathematical induction, for example, would go.[9]

Some might still argue that there are different degrees or kinds of misapplications involved in using diagrams and linguistic forms of representation. The claim would be that diagrams are intrinsically easy to misapply. In the previous examples, a diagram drawn on paper contains nonessential properties by which we tend to be misled. On the other hand, a proof in a linguistic text rules out this kind of accidental mistake. Therefore, it is important for us to ask whether or not the misapplication of diagrams stems from the intrinsic nature of diagrams. If this is the case, then it will be quite reasonable not to use diagrams for a valid proof.

In the present work, I aim to answer the preceding question in the negative. That is, the misapplication of diagrams is not intrinsically related to the nature of diagrams. Accordingly, my goal is to show that the traditional worry about the misuse of diagrams does not justify the general prejudice against diagrams. The possibility of misusing diagrams is not legitimate evidence to support the conclusion that no diagram can be used for a "real" proof. Achieving this goal would call into serious question the strong prejudice against nonlinguistic representation systems in the history of logic and mathematics. The following work will show that diagrams are not inherently misleading. The opinion that diagrams cannot be used for a valid proof will turn out to be an unsupported and wrong one. To demonstrate the logical status of diagrams I will carry out a case study on Venn diagrams, a very simple diagrammatic system which has been much used in mathematics and in logic.

Venn diagrams are widely used to solve problems in set theory and to test the validity of syllogisms in logic. In elementary school one is taught how to draw Venn diagrams for a problem, how to manipulate them, how to interpret the resulting diagrams, and so on. However, it is a fact

[9] Barwise and Etchemendy [4].

that Venn diagrams are not considered valid proofs, but heuristic tools for finding valid formal proofs.

Venn diagrams were invented by a nineteenth century logician, John Venn. Venn improved the earlier diagrams of Euler to form his own new diagrammatic method which he believed to be "in complete correspondence and harmony" with Boolean algebra. Many people have shared this belief. However, interestingly enough, nobody, including Venn himself, has tried to prove that this belief about his new diagrammatic system is true. Kneale and Kneale express a standard view:

> In his *Symbolic Logic* of 1881 J. Venn, another admirer of Boole, used diagrams of overlapping regions (i.e., topological models) to illustrate relations between classes or the the truth-conditions of propositions.[10]

Unless we clarify the meaning of "illustrate," this passage is not enough to convince us that drawing Venn diagrams is as valid as writing a proof in a first-order language. Martin Gardner, who explored the history of diagrams, was very enthusiastic about Venn diagrams. However, he expressed his enthusiasm in the following assertion, which, however, is not a proof:

> ...shortly after Boole laid the foundations for an algebraic notation, John Venn came forth with an ingenious improvement on Euler's circles. The result was a diagrammatic method so perfectly isomorphic with the Boolean algebra, and picturing the structure of class logic with such visual clarity, that even a non-mathematically minded philosopher could "see" what the new logic was all about.[11]

Many of us consider the Venn diagrammatic method to be a successful attempt at a topological illustration of Boolean algebra. It is an interesting question why everyone, including Venn, has stopped at this evaluation of Venn diagrams. Until now there has been no attempt to prove that there is an isomorphism between Boolean algebra and Venn diagrams. Nor has anyone tried to show that Venn diagrams form a sound and complete system independent of Boolean algebra. Achieving the latter task will be one goal of this work.

There seem to be two main reasons for this neglect of the exact status of Venn diagrams. One is an historical reason: The proof of the soundness and the completeness of representation systems had to wait for the development of semantics. Even if Venn had wanted to prove that his new diagrammatic system was a sound and complete system,

[10]Kneale and Kneale [14], p. 420.
[11]Gardner [10], p. 39.

he would not have had the appropriate tools to prove it. However, this is not the only reason why we do not get full credit for drawing Venn diagrams as a valid proof. The truth is that even after semantics was introduced into logic, no one proved that Venn diagrams form a sound and complete system.

The other reason for not inquiring into the status of Venn diagrams originates in the general prejudice against visualization in the mathematical tradition, which I have already addressed at length. With the corresponding bias for linguistic representation systems, little attempt has been made to analyze any nonlinguistic representation system, despite the fact that many forms of visualization are used to help our reasoning. Venn started the chapter of his *Symbolic Logic* where he introduced his new diagrammatic method with the following passage:

> The majority of modern logical treatises make at any rate occasional appeal to diagrammatic aid, in order to give sensible illustration of the relations of terms and propositions to each other.[12]

As this passage shows, Venn himself seemed to be satisfied with the role of diagrammatic methods as an aid to which one makes an occasional appeal. He did not give any clear definition of "illustration." However, he seemed to have more faith in the diagrammatic method than most other logicians have had. Even though he could not prove that his diagrammatic method is sound and complete, the following passage reveals his strong belief that a diagrammatic system, if correct, is as good as a correct symbolic system:

> Of course, we must positively insist that our diagrammatic scheme and our purely symbolic scheme shall be in complete correspondence and harmony with each other. ... But symbolic and diagrammatic systems are to some extent artificial, and they ought to be constructed as to work in perfect harmony together. This merit, so far as it goes, seems at any rate secured on the plan above.[13]

The present work aims to turn Venn's strong belief about the complete correspondence and perfect harmony between symbolic and diagrammatic systems into a demonstrated fact. This goal will be accomplished in the following way: Venn diagrams are presented as a standard formal representation system with its own syntax and semantics, and the system is proven to be sound and complete.

Before proceeding further, let me discuss a possible objection against the way I will accomplish the goal of this work. Some might think that

[12]Venn [27], p. 110.
[13]Venn [27], p. 139.

the division between syntax and semantics, the soundness proof, and the completeness proof belong to symbolic systems only. These objectors would argue that the way I achieve my goal defeats the purpose of my project – to give an independent (not just heuristic) status to diagrams – because I am using the tools that are properly applied to symbolic systems only. However, these objectors would be wrong. A division between syntax and semantics is not intrinsic to symbolic systems only. For *any* representation system, whether it is symbolic or visual, we can discuss two levels, a syntactic level and a semantic level. This is what representation means. The same argument goes for the soundness and the completeness proofs. What a soundness theorem says is that whenever we can obtain a unit of a system (say, x) from the given units of the system (say, $y_1, \ldots$) in the system, x is a logical consequence of $y_1, \ldots$. Units x, $y_1, \ldots$ do not have to be linguistic at all. If a representation system is not known to be sound, we will never be able to consider using the system for a valid proof, whether the system is linguistic or not. What a completeness theorem says is that whenever a unit of the system (say, x) follows from the given units of the system (say, $y_1, \ldots$), x is obtainable from $y_1, \ldots$ in this system. There is no constraint that x, $y_1, \ldots$ should be symbolic at all. I do not claim that every representation system should be proven to be complete. Some incomplete systems can still be used. In any case, syntax, semantics, soundness, and completeness are aspects of representation systems, but these concepts themselves are not tied up with any specific kind of system.

Before presenting Venn diagrams as a formal system, I will introduce in the next chapter three stages in the development of Venn diagrams, that is, Euler's, Venn's, and Peirce's systems. These preliminaries will serve two purposes. One is to give a brief historical sketch for Venn diagrams. The other, more importantly, is to show where each of the previous systems started and stopped, and to clarify where the novelty of my work on Venn diagrams resides.

In the main part of the present work, I will show that the Venn diagrammatic method, after a slight modification, can be proven to be sound and complete. I call this slightly modified version of Venn's original method Venn-I. The syntax and a semantic analysis for this system will be presented in Chapter 3. At the end of Chapter 3, I will prove the soundness and the completeness of Venn-I. The soundness of Venn-I assures us that if we manipulate Venn diagrams only according to our transformation rules, then this diagrammatic method will not mislead us. This assurance shows that one of the main worries of mathematicians and logicians about diagrams is not justified; rather, the use of diagrams does not tend to yield fallacies compared with the use of symbolic systems. Just as failure to apply a sound inference rule leads to fallacies in any linguistic representation system, worrisome diagrammatic fallacies we discussed earlier do not originate in the nature of diagrams but in a lack of clarity about the legitimate manipulation of diagrams. The validity of the transformation rules of Venn-I makes it clear that logical soundness can be obtained in a nonlinguistic form of reasoning. Therefore, the semantic analysis of a Venn system is interesting because it shows that the prejudice against visualization is not well grounded.

The analysis of Venn-I will lead to interesting issues which have their analogues in other deductive systems. Interestingly, Venn-I, whose primitive objects are diagrammatic, not linguistic, casts these issues in a different light than is usual with linguistic representation systems such as first-order logic. This Venn-I system helps us to realize what we take for granted in other more familiar deductive systems.

The Venn-I system is quite limited in representing information. For example, this system does not have a general way of conveying disjunctive information. However, I do not suggest that this limit prevents us from using Venn-I. There are also many kinds of information that neither sentential logic nor first-order logic can convey, as mentioned earlier in this chapter. This fact does not prevent us from using either of these systems when we need to.

In Chapter 4, I extend Venn-I to give it more expressive power and call it Venn-II. Venn-II handles disjunctive information in a more general way than Venn-I. New formation rules are introduced to increase the expressiveness of the system, especially its ability to express disjunctive information. Accordingly, new transformation rules are needed in order to handle these new meaningful units. The semantics of this system is also formalized in terms of information-theoretic tools. After establishing the syntax and the semantics of Venn-II, I prove this system is also sound and complete. In Chapter 5, I show that Venn-II and a monadic first-order language are equivalent to each other.

As mentioned in the first paragraph of this chapter, there are other issues involving diagrammatic representation besides the formalization of a diagrammatic system. For example, the comparison between different kinds of representation systems (e.g., linguistic symbols and diagrams), the special aspects of diagrammatic knowledge representations, and so on, have been discussed by philosophers, psychologists and computer scientists. However, I strongly believe that real comparison can be done more substantially only after we consider diagrammatic methods in a more rigorous way than we have done traditionally. As long as we treat diagrams as secondary or as mere heuristic tools, we cannot make a real comparison between visual and symbolic systems. After being free from this unfair evaluation among different representation systems, we can raise another interesting issue: If a visual system is also a standard representation system with its own syntax and semantics, then what is a genuine difference, if any, between linguistic and visual systems? The answer to this question will be important not just as a theoretical curiosity but for practical purposes as well. If there is any intrinsic difference between diagrammatic and linguistic systems, then depending upon the nature of the information we want to represent, we can decide which kind of a system we want to adopt. All these topics are beyond the scope of the present work. In Chapter 6, I will deal with these issues in a limited way, suggesting several important features that lead to a distinction between linguistic and diagrammatic systems, using features of the two Venn systems by way of illustration.

2
Preliminaries

There has been some disagreement about when circles (or closed curves) began being used for representing classical syllogisms. They seem to have first been put to this use in the Middle Ages.[1] However, there seems to be agreement that it was Leonhrad Euler, in the eighteenth century, who proposed using circles to illustrate relations between classes. This diagrammatic method of Euler's was greatly improved by a nineteenth century logician John Venn. And in this century, it was Charles Peirce who made a great contribution to the further development of Venn diagrams.

This chapter explores the essence of Euler diagrams and their descendants, and will serve to prepare the reader for my approach to Venn diagrams presented in the following chapters. In each section, along with the main ideas of each system and its limits, I focus on how some of the main limits of one system are overcome by the following system. That is, the Venn system solves some of the main problems that the Euler system has. This improvement was significant enough to make necessary a distinction between Euler diagrams and Venn diagrams.[2] I will show that Peirce's revolutionary ideas about diagrams not only overcame some important defects of Venn diagrams but opened the way to a totally new horizon for logical diagrams. This last aspect will be discussed in detail in the third section. I will also point out where this new horizon stopped, and will claim that my approach to Venn diagrams (in Chapters 3 and 4) is the natural completion of these predecessors' incomplete projects.

Sun-Joo Shin, Peirce and the Logical Status of Diagrams, *History and Philosophy of Logic* 15 (1994): 45–68.

[1]Ramon Lull, *Ars Magna*, Lyons, 1517.

[2]However, some still use the name "Euler diagrams" for both systems. I will point out later that Peirce was one of them.

2.1 Euler diagrams

Euler introduced his circles in *Lettres à une princess d'Allemagne* (1761) to illustrate syllogistic reasoning. The four kinds of categorical sentences are represented in this system as follows:

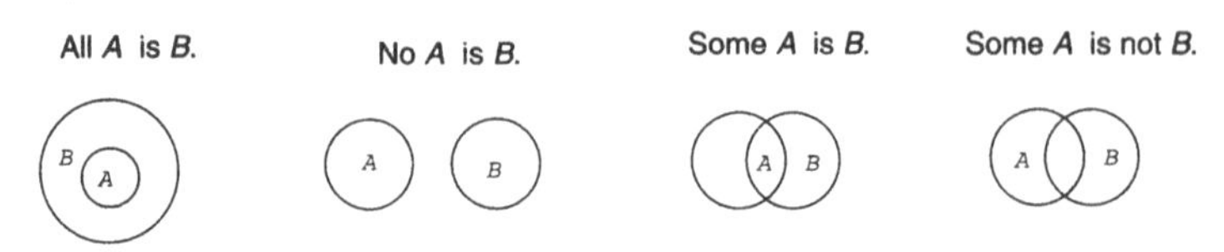

The two universal statements are represented in an intuitive way: If the circle with the letter "A" is included in the circle with "B," then the diagram represents the information that all A is B. If there is no overlapping part between two circles, then the diagram conveys the information that no A is B. However, the existential statements, "Some A is B" and "Some A is not B," are represented in a rather unclear way. Both diagrams mean to say the following: Something that belongs to A exists where letter "A" is written. Accordingly, where letter "A" is written turns out to be crucial in each representation. If we erase the letter "A" in each diagram, there is no distinction between these two diagrams.

This attempt is definitely different from other symbolic representation systems in that logical relations are illustrated geometrically. In this respect, Euler is not the first one who made this kind of attempt.[3] In the following passage, Martin Gardner points out a main difference between Euler diagrams and previous diagrammatic representations:

> Here [in Euler's *Lettres à une princess d'Allemagne*] for the first time we meet with a geometrical system that will not only represent class statements and syllogisms in a highly isomorphic manner, but also can be manipulated for the actual solution of problems in class logic.[4]

A main merit of the Euler system is correctly identified by Gardner as follows: We can manipulate Euler diagrams to demonstrate syllogistic reasoning. What does he mean by a geometrical system being manipulated to help us solve syllogistic reasoning problems? Consider two examples used by Peirce.[5]

Example 1 *Is this syllogism valid?*

All men are passionate.

[3] Gardner [10], pp. 30–1.
[4] Gardner [10], p. 31.
[5] Peirce [18], 4.350.

All saints are men.
Therefore, all saints are passionate.

Example 2 *Is this syllogism valid?*

No man is perfect.
But any saint is a man.
Hence, no saint is perfect.

The following diagrams are drawn for the premises, according to Euler:

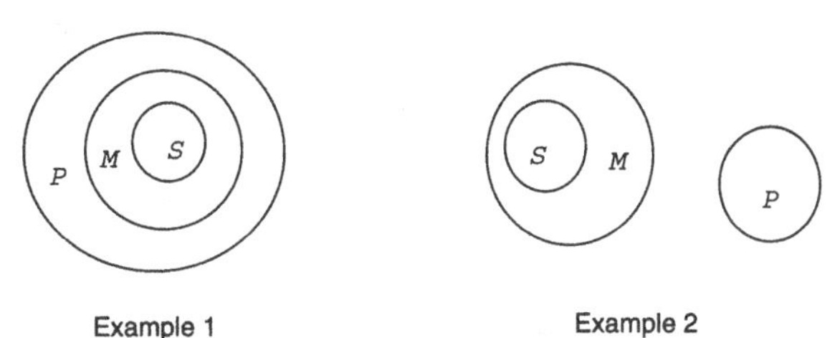

Example 1 Example 2

The first diagram shows that the circle for the saints is included in the circle for the passionate. That is, all saints are passionate. The second diagram shows that there is no overlapping part between the circle for the saints and the circle for the perfect. That is, no saint is perfect. In this way, we correctly solve the problems. This is what Gardner means when he says that we can manipulate Euler diagrams to solve problems. This process is definitely novel in that the system can handle deductive reasoning, although, as we will see, it is far from being perfect.

Many people have agreed that the following is one main defect of the Euler system: This system does not exhaust all the possible relations between two terms. Peirce cited a list of the possible different ways in which two terms are related.[6] Among these ways, the Euler system does not represent, for example, the following propositions involving two terms, S and P: Everything is either S or P; everything is both S and P; nothing is S but everything is P; and so on.[7] This may seem to be a major drawback, but in the following William Kneale and Martha Kneale point out that exhausting all possible relations was not Euler's concern at all:

> In Euler's version of logic the chief role of spatial figures is to make the principles of syllogistic seem intuitively evident, and there is no attempt to survey all possible relations of extensions;...[8]

[6]Peirce attributed this list to Mrs. Franklin. Refer to Peirce [18], 4.356.
[7]Peirce [18], 4.356.
[8]Kneale and Kneale [14], p. 350.

According to this interpretation, the only propositions Euler aimed to represent are those four categorical ones. If so, the defect in question might not be as fatal as one might think. After all, if needed, we could extend this system to include the representations of all possible relations. Therefore, it might be more fair to limit ourselves to Euler's representations of the four propositions and ask what problems this system has. I ran into three related obstacles when I tried to use these diagrams for syllogistic reasoning.

The first difficulty lies in an ambiguity of the following diagram for a negative existential proposition, "Some *A* is not *B*."

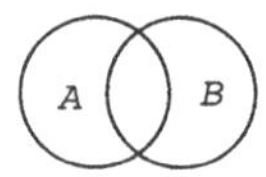

The following is Euler's explanation of how this diagram represents the proposition "Some *A* is not *B*": "A part of concept *A* is located outside of concept *B*."[9] The problem is that this passage also holds for the following diagram, which is supposed to represent a different proposition, "Some *B* is not *A*":

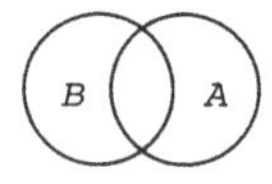

This diagram shows not only that a part of *A* is not included in *B* but that a part of *B* is not included in *A* either. Then, the question is whether this diagram represents the proposition "Some *B* is not *A*" or the proposition "Some *A* is not *B*." That is, there is a risk that this system might be ambiguous, which is a fatal drawback in a representation system.

The second defect is found in Euler's representation of contradictory propositions. For example, "All *A* is *B*" and "Some *A* is not *B*" convey contradictory information. In the case of Boolean algebra, the two statements $A\overline{B} = 0$ and $A\overline{B} > 0$ show the contradiction in an obvious way. The same goes for $AB = 0$ and $AB > 0$. However, the Euler system does not express the contradiction in an obvious way. Let us compare these two diagrams to see how a contradiction is represented in this system:

[9]Euler [9], p. 119

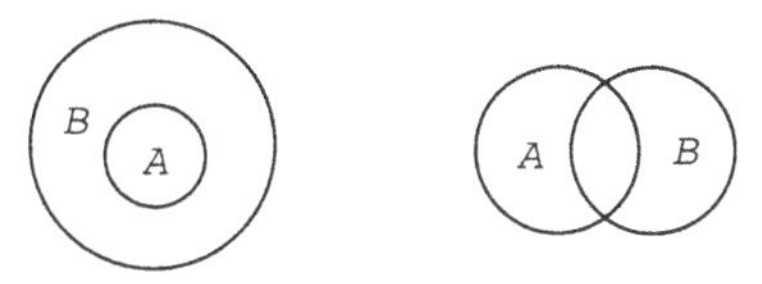

Some might say that these two diagrams convey a contradiction since *A* is included in *B* in the left figure while part of *A* is not included in *B* in the right one. However, in a diagram for "No *A* is *B*," *A* is not included in *B* and in a diagram for "Some *B* is not *A*" part of *A* is not included in *B* either, as follows:

But we do not want to say that either of these diagrams conveys information that contradicts the information that the diagram for "All *A* is *B*" conveys. Some might object to this criticism of Euler's representation of contradiction, saying that a system does not have to express contradiction in an obvious way. However, I think this defect is related to the most serious problem of this system which I will discuss next.

The third problem is the difficulty in combining more than one piece of information. Let's compare the following example with the previous two examples, Examples 1 and 2.

Example 3 *Can we illustrate the validity of the following syllogism?*

> All *A* are *B*.
> Some *A* are *C*.
> Therefore, some *B* are *C*.

The following are the diagrams for each premise:

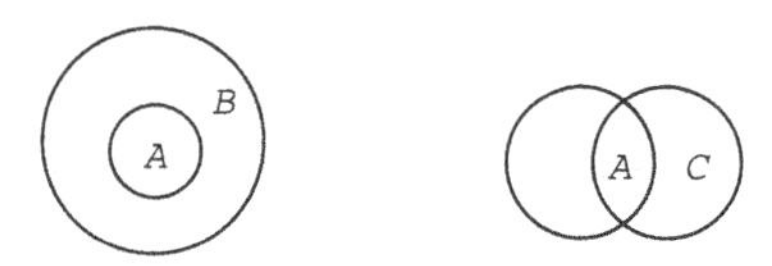

We need to combine these two diagrams. Otherwise, we have no way to know the relation between *B* and *C*, which the conclusion is concerned with. However, these two diagrams cannot be manipulated into one, unlike the diagrams of Examples 1 and 2. It should be recalled that Gardner considered manipulability the best merit of this system. Unfortunately, this merit does not seem to work for more general cases.

Some might think that this incapability has something to do with Euler's unclear representation of existential statements, which we discussed as the first defect. But, we cannot attribute this serious fault to the representations of existential propositions, because the same kind of problem arises when we have only universal statements as follows:

Example 4 *Can we unify the following two diagrams into one?*

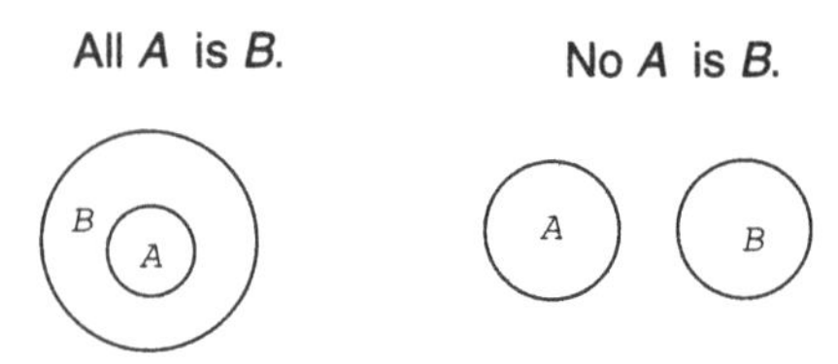

We cannot combine these two diagrams in this example (which has only universal categorical statements) either. What is worse in this case is that both diagrams look incompatible with each other, whereas both propositions can be true at the same time if there is no A.

The first problem concerning the representation of existential statements could be easily solved. For example, the order of the letters can be considered an important representing fact. Then, the following diagrams represent different propositions:

The second dissatisfaction about the representation of contradiction might not be crucial since contradiction does not have to be obviously represented. But, the third one – compatible pieces of information are not always representable in one diagram – is a crucial drawback. If we cannot combine more than one diagram freely and we do not know how to transform a diagram, then the system is severely limited in handling syllogistic deductive reasoning.

2.2 Venn diagrams

The realization of the faults of Euler diagrams leads us to the appreciation of Venn diagrams. Venn finds the Euler system too strict in the sense that this diagrammatic method is not general enough to represent more than one possible relation between two classes in the same diagram.[10] This is closely related to the third defect discussed in the previous section, that is, the difficulty in combining diagrams. This difficulty can be interpreted as one of the inevitable results of this strictness. Venn succinctly sums up his criticism against the Euler system in the following passage:

> The weak point in this [Euler diagrams], and in all similar schemes, consists in the fact that they only illustrate in strictness the actual relation of classes to each other, rather than the imperfect knowledge of these relations which we may possess, or may wish to convey by means of the proposition.[11]

In the Euler system, each diagram represents facts in an overly restricted way. There is no room for representing partial information. If we could represent partial information in a diagram, then we could add new information in the same diagram. As we have seen, this possibility is lacking in the Euler system. The reason why Euler diagrams fail in representing this kind of possible compatibility between information is simple: This system does not have a way to represent uncertainty of information. Whenever circles are drawn, they exclude the possibility of representing other compatible information.

Venn's new system was to overcome this defect. He improved Euler's system in such a way that in one diagram all possible relations between classes might be represented. First, we draw two circles in the following way:[12]

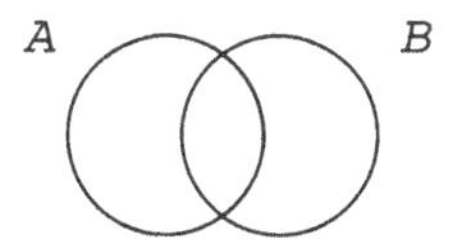

According to this new system, this diagram does not convey any information about the relation between two classes. Venn calls this a primary diagram. It should be noticed that the use of a primary diagram leads to a fundamental difference between Euler and Venn diagrams. There is no primary diagram in the Euler system. This is why we need to draw a new diagram for each piece of information, rather than being able to add new information to an existing diagram. Each compartment of Venn's

[10] Notice that this criticism is different from the criticism that many have raised and I mentioned in the previous section (i.e., the criticism that the Euler system does not represent all the possible relations between two terms). Venn is concerned with representing more than one relation in *one* diagram.

[11] Venn [27], p. 510.

[12] Venn [27], p. 114.

primary diagram represents a possible combination of class A, class B, class $\overline{A}$, and class $\overline{B}$. Influenced by Boolean algebra, Venn expresses each compartment in the following way:

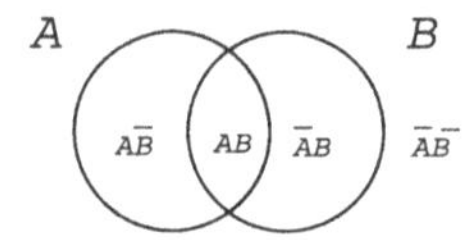

The next step is to represent certain information about the relations between two classes in this diagram. The way two circles overlap does not represent any specific relation between the two classes, unlike with the Euler system. Depending on which class an assertion is concerned with, we find an appropriate area in a primary diagram and express the information in it. Obviously, we need extra syntactic devices in this system. Venn adopts a visual object, shading, to place in an appropriate compartment:

> What we do then, is to ascertain what combinations or classes are negatived by any given proposition, and proceed to put some kind of mark against these in the diagram. For this purpose the most effective means is just to shade them out.[13]

For example, proposition "All A is B" conveys the information that nothing exists in class $A\overline{B}$. What Venn does is to find the area whose name is $A\overline{B}$ and put a shading in that region, as follows:

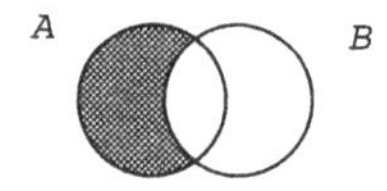

What makes this representation different from Euler's representation of "All A is B"? Some might say that Euler's representation is more intuitive than Venn's. The following might be why: In Euler's, the inclusion relation between two classes is represented in a more evident way by means of the inclusion of one circle within the other. Also, for Venn's representation, a new syntactic device, shading, needs to be explained. Nevertheless, Venn's representation has a strong advantage over Euler's in that the previous diagram leaves room for other information. Suppose we want to add another piece of information, say, "No A is B", to the foregoing information. What we do is put a shading in area AB.

[13]Venn [27], p. 122.

The following process shows how to combine more than one piece of information.

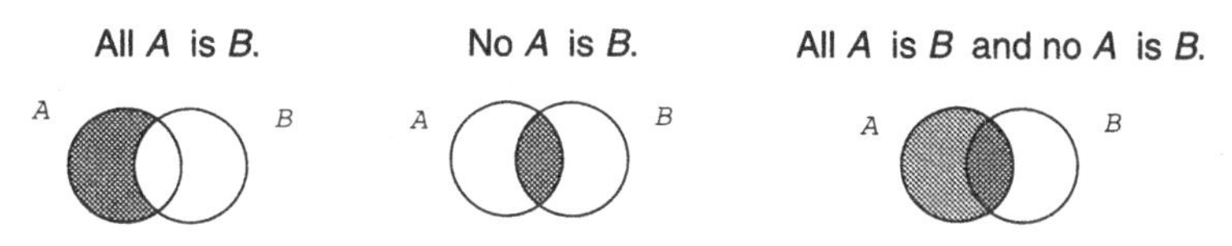

This is where the Euler system failed, as shown earlier in Example 4. This new system allows one to combine these two pieces of information without any difficulty. After demonstrating how to draw his new diagrams, Venn makes the following comparison of his diagrams and Euler's:

> How widely different this plan is from that of the old-fashioned Eulerian diagrams will be readily seen. One great advantage consists in the ready way in which it lends itself to the representation of successive increments of knowledge as one proposition after another is taken into account, instead of demanding that we should endeavor to represent the net result of them all at a stroke.[14]

We can also see that a common complaint against the Euler system (i.e., it does not exhaust the list of all the possible relations between two terms) is largely defused. For example, the proposition "Anything which is either *A* or *B* is both *A* and *B*" is represented in the following way:

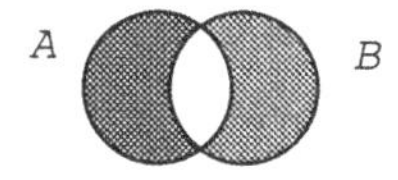

When Venn overcame the strictness of Euler diagrams, he eliminated other defects already noted. The main innovation of Venn's system over Euler's may be summarized as follows: Venn's system represents the different possible combinations among four classes (for two given terms) by means of different compartments in one diagram. Accordingly, we may always add more information without drawing a new diagram as long as that new information is concerned with relations among the classes drawn on that diagram.

The question remains how Venn handles existential propositions, which caused trouble for Euler. Surprisingly, Venn does not deal with this question. In that sense, we cannot tell whether Venn overcame all of Euler's problems or not, since some of my criticisms of the Euler system involve the representations of existential statements. However, we can imagine that Venn could have introduced another kind of visual object

[14]Venn [27], p. 123.

that can be put in an appropriate region with existential import. This is exactly what Peirce did about 20 years later.[15] We will discuss Peirce's idea in detail in the next section.

Before moving to the next phase in the development of Venn diagrams, I want to point out Venn's confusion between syntax and semantics, which will be a main issue in the discussion of Peirce as well. This confusion is well revealed in the following question:

> The best way of introducing this question [how Venn's diagrammatic scheme can be worked so as to represent propositions] will be to enquire a little more strictly whether it is really *classes* that we thus represent, or merely *compartments* into which classes may be put?[16]

From our point of view, this question itself does not make sense, since the two alternatives Venn suggests, that is, *classes* and *compartments*, do not belong to the same category. Compartments represent classes in a diagram. That is, a compartment with a shading is a *representing* fact, whereas a class without any member is a *represented* fact. This passage clearly shows that we cannot expect Venn to perform a semantic analysis of his diagrammatic system. He definitely did not have appropriate tools to carry on this project. We need to wait for the development of semantics in the history of logic.

2.3 Venn–Peirce diagrams

Charles S. Peirce invented a system of logical diagrams which he called "Existential Graphs" in 1896. Before introducing his own graph system, Peirce discussed Euler diagrams (or Venn diagrams)[17] at great length. I want to emphasize two important aspects of this discussion. One is the five faults of Euler diagrams Peirce suggested. The other is his rules of transformation. In the following subsections, each of these topics will be taken up in turn.

[15] Peirce [18], 4.357.
[16] Venn [27], p. 119.
[17] Peirce calls Venn diagrams (in our sense) "Euler diagrams" throughout his writing. Peirce also named the section "Of Euler diagrams." It is true that Euler first introduced circles to carry out syllogistic reasoning. However, I will show in the following subsection that the name "Venn diagrams" is more appropriate than the name "Euler diagrams" in this context.

2.3.1 Modification of Venn diagrams

Peirce starts his discussion with Euler diagrams, but immediately accepts Venn's improvement on Euler diagrams. This acceptance comes from the realization that the Euler system has no means to represent all the possible ways in which two terms may be related. This realization is the first of the five defects that Peirce raises against Euler's system. From that point on, what Peirce discusses in the section "Of Euler diagrams" is actually Venn's substantially modified version. Accordingly, in this section I replace Peirce's use of the term "Euler diagrams" with "Venn diagrams." After adopting the Venn system, the following four problems are left to be solved:

(i) The Venn system "*cannot affirm the existence* of any description of an object."[18] Venn never tells us how to represent existential statements. As seen in the previous section, Venn's shading is capable of representing universal statements only.

(ii) The system "affords no means of expressing a knowledge that one or another of several alternative statements of things occurs."[19] That is, the Venn system is not able to express disjunctive information.

(iii) "It cannot express enumerations, statistical facts, measurements, or probabilities."[20] In this respect, a standard first-order language containing existential and universal quantifiers has the same kind of problem.

(iv) "It does not extend to the logic of relatives."[21] That is, relations cannot be represented in this system.

This dissatisfaction leads Peirce to two important results. First, he modifies Venn diagrams to overcome (i) and (ii). Second, he invents his own graph systems, realizing that the last problem cannot be solved in the Venn system. For my purpose, the first result is extremely important. The Venn systems I will discuss in the following pages are a partial adoption of Peirce's modification of Venn diagrams, which Merrilee Salmon also uses in her introductory logic book.[22] Let us consider in detail how Peirce extends or modifies Venn diagrams.

[18]Peirce [18], 4.356.
[19]Peirce [18], 4.356.
[20]Peirce [18], 4.356.
[21]Peirce [18], 4.356.
[22]Salmon [22], pp. 275–92.

In order to represent existential statements, Peirce introduces the character "x" into the system. This extended system allows us to represent the following propositions in the following way:[23]

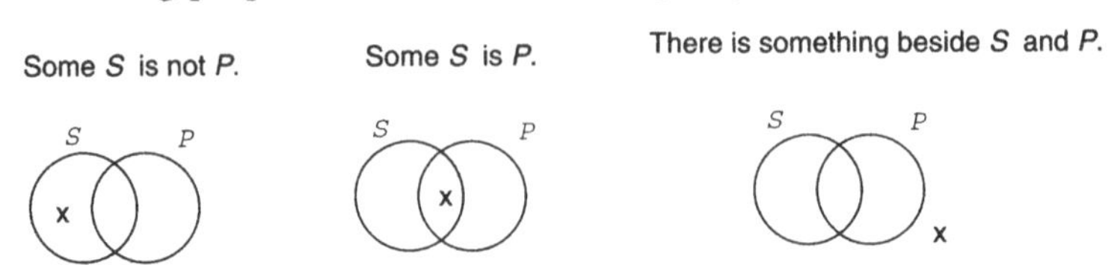

In order to represent disjunctive information, Peirce takes two steps. First, he replaces Venn's shading with the character "o." This modification is illustrated in the following:

Second, he suggests the following rule (he calls it a rule): "Connected assertions are made alternatively."[24] Peirce introduced a syntactic device to represent disjunctive information. The device is a line, which connects marks ("o" or "x").[25] For example, the rightmost diagram in the following conveys the information that is a disjunction of the information that the leftmost diagram conveys and the information that the middle one conveys:

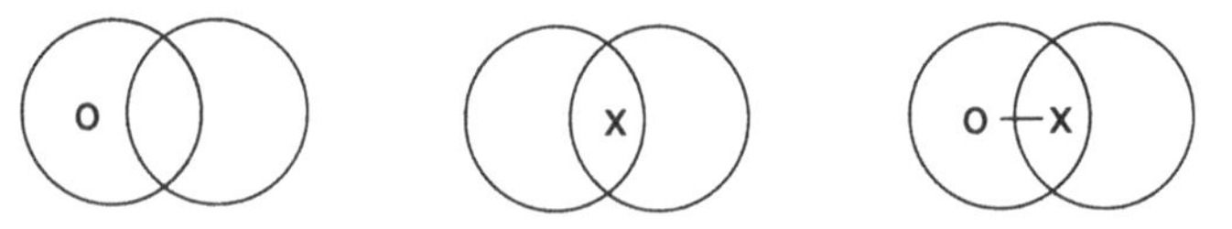

However, in more complicated cases, this way of representing alternative information is not so simple. In the following example, the rightmost diagram conveys the disjunction of the information the left two diagrams represent:

[23]Peirce [18], 4.359.
[24]Peirce [18], 4.360.
[25]Even though Peirce does not say why he uses "o" instead of a shading, the difficulty of connecting shadings (for disjunctive information), I believe, is one of the main reasons for this modification.

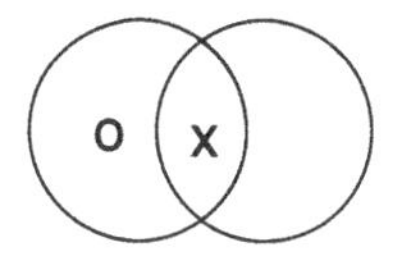

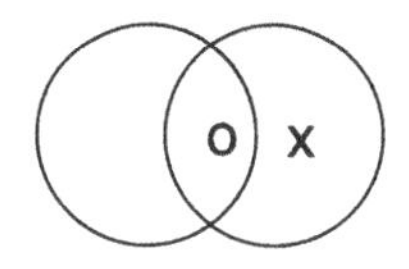

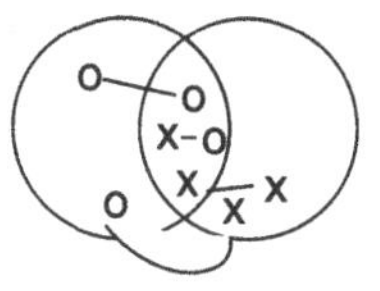

The rightmost diagram does not seem to have much visual power. At this point, Peirce's idea of connecting signs for disjunctive information seems to start undermining the visual effect that the original Venn method possesses. Peirce's suggestions to overcome (ii), the problem of disjunction, yield the following result: The modified system increases the power of expression, but loses the visual effects of a diagrammatic system. Peirce himself admits the complexity of his suggestion for connecting characters, and suggests an alternative way:

It is only disjunctions of conjunctions that cause some inconvenience; ...It is merely that there is a greater complexity in the expression than is essential to the meaning. There is, however, a very easy and very useful way of avoiding this. It is to draw an Euler's [Venn's] Diagram of Euler's [Venn's] Diagrams each surrounded by a circle to represent its Universe of Hypothesis. There will be no need of connecting lines in the enclosing diagrams, it being understood that its compartments contain the several possible cases.[26]

The basic idea behind this suggestion is very similar to the idea behind the disjunctive normal form of symbolic logic. Each compartment of a diagram conveys only conjunctive information and the relations among the compartments are disjunctive. According to this modification, the previous rightmost, complicated-looking diagram may be changed into the following diagram:

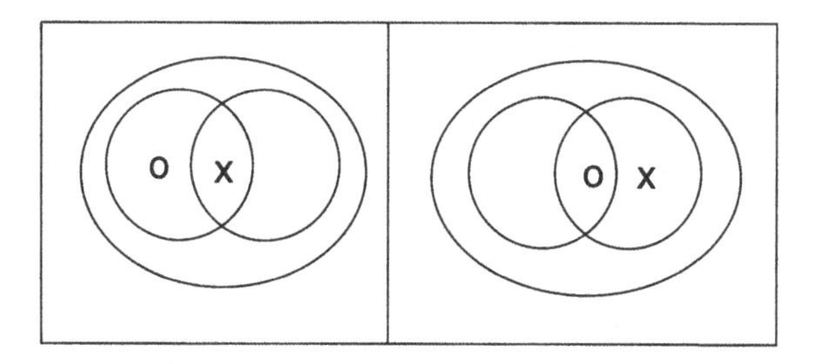

My presentation of Venn diagrams keeps to Venn's original system, and accepts only those extensions of Peirce's that may consistently be applied to the Venn system. That is, I will use a shading (not "o"), and

[26]Peirce [18], 4.365.

three new syntactic devices, "x," a connecting line between x's (for Venn-I in Chapter 3), and a connecting line between diagrams (for Venn-II in Chapter 4).

2.3.2 *Rules of transformation*

Another important contribution of Peirce's to Venn diagrams is to introduce rules of transformation into this system. I want to focus on three aspects of Peirce's discussion of rules in the following subsections in turn. (i) I will discuss the purpose of Peirce introducing rules. (ii) I will examine each of Peirce's rules. (iii) I will point out that Peirce's unclear distinction between syntax and semantics of a representation system prevents him from achieving his original goal.

Peirce's intuition about rules

Peirce makes the following interesting comments on rules just before he enumerates his rules:

> "Rules" is here used in the sense in which we speak of the "rules" of algebra; that is, as a permission under strictly defined conditions. (Footnote: This curious use of the word *Rule* is doubtless derived from the use of the word in Vulgur Arithmetic, where it signifies a method of computation adapted to a particular class of problems; as the Rule of Three, the Rule of Alligation, the Rule of False, the Rule of Fellowship, the Rule of Tare and Tret, the Rule of Coss. Here the Rule is a body of directions for performing an operation successfully. But when we speak of the Rule of Transposition, the directions are so simple, that the Rule becomes principally a permission.)[27]

Peirce was probably the first person that discussed the rules of transformation in a diagrammatic system. The quoted passage also shows that Peirce had a quite clear idea of what these rules are supposed to do. As we have rules in algebra, we may have rules for diagrams. The rules of algebra tell us what we are permitted to do and what we are not. Obviously, Peirce wants to have transformation rules of diagrams that tell us what we are permitted to do or what we are not. What does Peirce mean by a "permission"? Why do we need a "permission" in order to use diagrams? How did either Euler or Venn use diagrams without these rules? Let me consider several examples of syllogistic reasoning to answer these questions:

[27] Peirce [18], 4.361

Example 5 [28]

No Y is Z.
All X is Y.
Therefore, no X is Z.

The following diagrams are the illustrations of this syllogistic reasoning by means of the three diagrammatic systems in question:

Fig.1

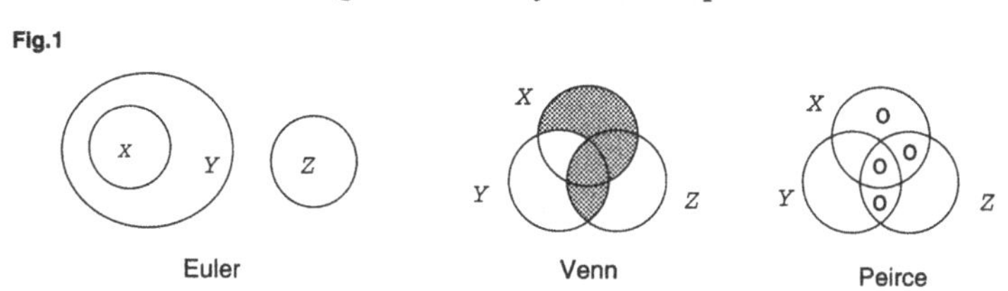

It has been believed that each of these diagrams (which represents the information conveyed by the two premises) *contains* the information conveyed in the conclusion, since the following diagrams, respectively, are included in an obvious way in the previous diagrams:

Fig.2

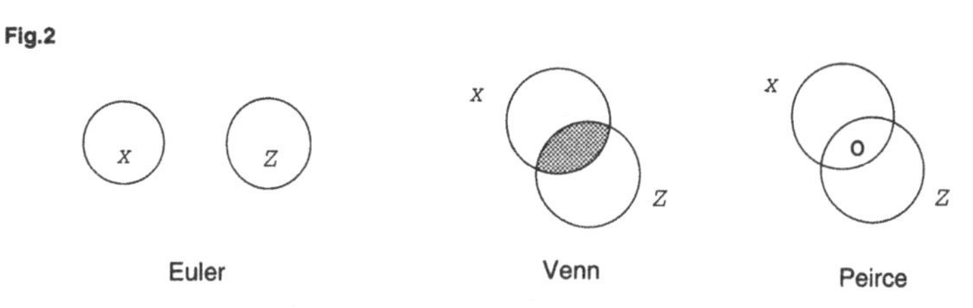

Accordingly, we believe that Euler diagrams, Venn diagrams, and Venn–Peirce diagrams illustrate this syllogistic reasoning. However, the concept "contain" needs further explanation. Suppose that the two premises of the syllogistic reasoning in Example 5 are represented in a first-order language in the following way:

$$\forall x(Yx \rightarrow \neg Zx).$$
$$\forall x(Xx \rightarrow Yx).$$

Is first-order sentence $\forall x(Xx \rightarrow \neg Zx)$ included in the preceding first-order sentences? No, at least not in the same sense as a diagram is included in another diagram in the previous way (cf. Fig. 1 and Fig. 2). However, in first-order logic we can prove that $\forall x(Xx \rightarrow \neg Zx)$ is a logical consequence of the two sentences, that is, $\forall x(Yx \rightarrow \neg Zx)$ and

[28] Venn uses this example for the comparison between Euler's and his own system, in Venn [27], pp. 125–6.

$\forall x(Xx \rightarrow Yx)$, and accordingly we say that the information conveyed in the sentences $\forall x(Yx \rightarrow \neg Zx)$ and $\forall x(Xx \rightarrow Yx)$ contains the information conveyed in the conclusion. Therefore, we know that first-order logic captures the valid syllogistic reasoning in Example 5. An important question is how this logical consequence relation is proven in first-order logic, despite the fact that the inclusion relationship is not as obvious as in the case of diagrams. Inference rules are used to derive a sentence we want from the given sentences. Some inference rules are clear formalizations which allow us to derive an included sentence from the given sentence. For example, in the following, the "$\wedge$" elimination rule allows us to derive sentence (2) from sentence (1):

$$\forall x Yx \wedge \forall x Xx. \tag{1}$$

$$\forall x Yx. \tag{2}$$

We might want to say that sentence (2) is included in sentence (1), just as we have thought that the diagrams in Fig. 2 are included in the corresponding diagrams in Fig. 1. Suppose we have some kind of elimination rule (which corresponds to the "$\wedge$" elimination rule in first-order logic) in each diagrammatic system. Then, the diagrams of Fig. 1 would be transformed into the diagrams of Fig. 2.[29]

One might think that this kind of rule is not needed, since it is visually obvious that one diagram is included in another. However, as we have seen in the case of first-order sentences (as in Example 5), sometimes it is not visually obvious that one sentence is included in another. This is where we rely on inference rules. Does this case (i.e., the nonobvious containment relation among diagrams) also happen in diagrammatic systems? The following example is one such case:

Example 6 *We want to know whether the following reasoning is valid or not.*

All X is Y.

No X is Y.

Therefore, no X is Z.

The following diagrams would be drawn for the premises in each system:

[29] For now, we just suppose that we have a well-specified rule for each system so that the rule correctly eliminates parts of a diagram.

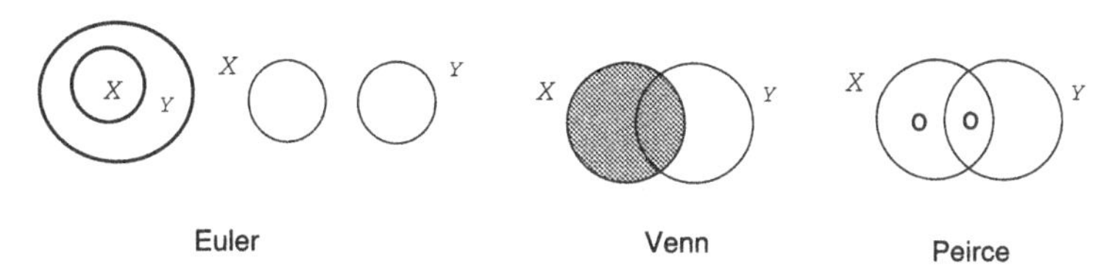

The trouble is that none of these diagrams has a circle for Z. Accordingly, the following diagrams (which represent the conclusion) cannot be included in the previous diagrams:

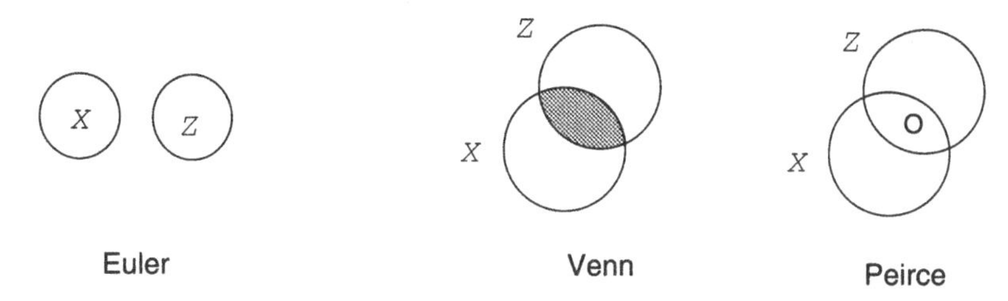

This example clearly shows that a visual inclusion relation is far from being sufficient to define the consequence relation. If we want any of these diagrammatic systems to illustrate the valid reasoning of Example 6, we should be permitted to draw a new circle for Z. That is, we need rules that tell us what we are allowed to do in a system or what we are not. Peirce's passage about the necessity of rules or "permissions" makes this point, using an analogy to algebra. Venn did not face this problem when he introduced his system. I suspect that what Venn had in mind is a very simple and obvious inclusion relation among diagrams. Accordingly, Venn's original system could not illustrate the validity of the reasoning in Example 6. When the Venn system is extended into the Venn–Peirce system, this need is more increased.[30] The following example can be illustrated only after we adopt Peirce's extended system:

Example 7

> No A is B.
> Some B is C.
> Therefore, some C is not A.

The premises are represented in the left diagram and the conclusion in the right one:

[30]It is because there are more syntactic devices in the latter system, and accordingly, the containment-relation becomes less visually obvious.

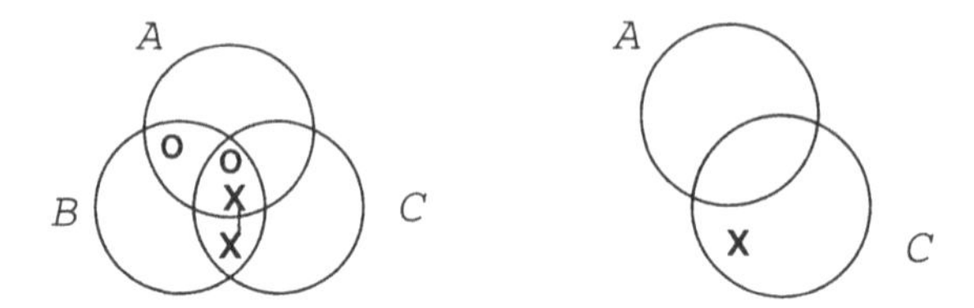

It is not clear whether the right diagram is included in the left diagram. If we want this system to be able to illustrate this valid reasoning, we should be able to show that the information represented in the right diagram follows from the information represented in the left diagram. Obviously, Peirce realizes this need very well. Peirce's intuition about rules, which he never states directly, might be summarized in the following way, which I will refer to as "PIR":[31]

(PIR) If we have transformation rules that correctly guide us on how to manipulate diagrams, then we get (all and)[32] only diagrams that are consequences of the given diagrams.

Whether Peirce himself realized it or not, this insight explains a fundamental difference between Peirce's approach to diagrams and any other previous attempts at the use of diagrams. PIR, I believe, led Peirce to invent his Existential Graphs, which are the first systematic and rigorous project for diagrammatic reasoning. My work on Venn systems definitely has been inspired by this implicit insight of Peirce's. Quite surprisingly, there seems to be nothing in the literature on diagrams, on Peirce's logic, or on Peirce's graph systems that shows an appreciation of this intuitive idea behind Peirce's rules of transformation. I will try to understand why many of us have missed the importance of PIR, after examining the transformation rules in detail.

Peirce's rules

In [18], 4.362, Peirce introduces six rules of transformation in a rather loose way. In this subsection, I aim to explain these rules one by one through examples. At the same time, we want to check whether each rule fulfills the goal behind PIR.

Rule 1: We may erase any entire sign ("o," "x," or a connected body).

[31] This is short for "Peirce's intuition about rules."

[32] Peirce does not think that his rules are complete, but believes that we can have a complete set of rules.

This rule permits us to transform the diagram on the left to the diagram on the right:

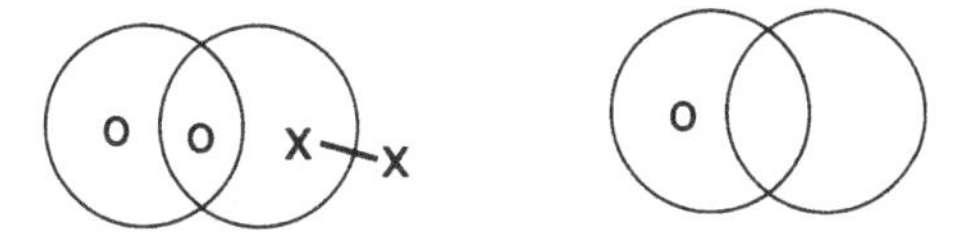

Intuitively this rule looks valid, since erasing a sign is erasing information from the given information. If the given information is true, whatever information we extract from it by erasing its part must also be true. The conjunction-symbol-elimination rule of symbolic logic does the same job as this rule.

Rule 2: We may connect a character to any existing character.

When we recall that Peirce introduced a connecting line in order to represent disjunctive information, this transformation seems to be quite intuitive. For example, a system will be able to demonstrate that the following inference is valid:

> All *A* is *B*.
> Therefore, all *A* is *B* or some *A* is *B*.

The application of rule 2 allows us to transform the left diagram (for the premise) to the right one (for the conclusion):

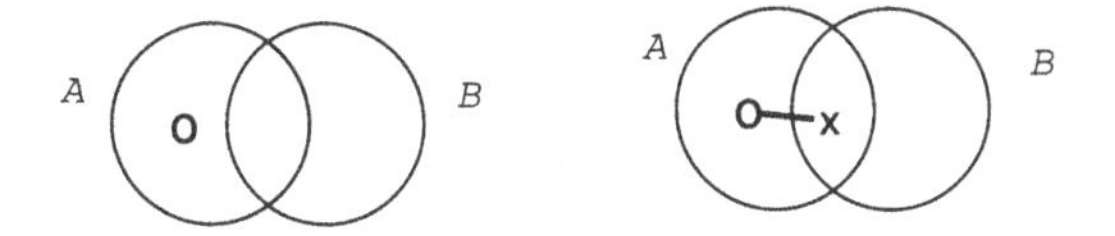

This rule is quite similar to the disjunction-introduction rule in symbolic logic.

Rule 3: Any assertion which could permissively be written, if there were no other assertion, can be written at any time, detachedly.

What does Peirce mean by "assertion which could permissively be written?" This should mean something other than what he means when he says that rules permit us to perform certain manipulations. Otherwise, this rule would be circular in that we need rules for permission but this rule is assuming certain permissions. I would take this phrase to mean assertion that is provided as a new piece of information. That is, rule 3 allows us to put in a new piece of information. For example, in the following, after having the diagram on the left, we want to add more information, say, "Some B is C." Then, we may write down two connected x's by this rule. This is the way we get the diagram on the right side:

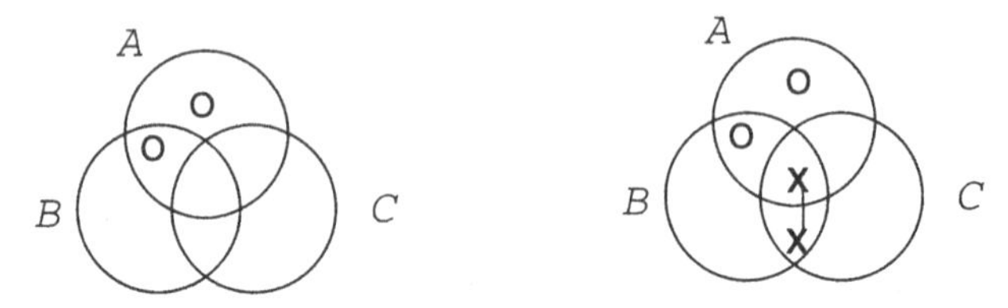

Rule 4: When more than one character is written in the same compartment, we may transform a diagram in several ways as follows:

(i) *Suppose that the characters in the same compartment are of the same kind of sign, whether attached or detached. Then, they are equivalent to one writing of it.*

(ii) *Suppose that they are of two different signs.*
(ii-1) Suppose that they are connected with one another. Then, we may erase them or insert them, since they are equivalent to no sign at all.
(ii-2) Suppose that they are detached from one another. Then, it constitutes an absurdity.

(iii) *Suppose that two different kinds of signs are connected with other signs, say, P and Q, respectively. Then, we may erase the two contrary signs and connect P and Q.*

By analyzing clause (i) and (ii) of this rule, I will show that Peirce does not make a clear distinction between syntax and semantics either.[33] This confusion leads him to several problematic treatments of diagrams, which I will discuss in the following paragraphs. More importantly, I claim that this confusion prevented Peirce from pursuing further steps for diagrammatic reasoning, which I will point out in the next subsection.

In clauses (i) and (ii-1), he mentions "being equivalent to." What does he mean by this? According to him, all the following diagrams are equivalent to each other:

[33] Recall Venn's confusion between syntax and semantics, which was discussed at the end of §2.2.

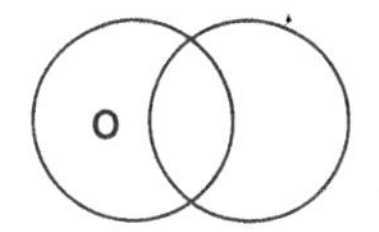

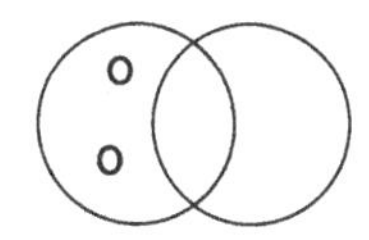

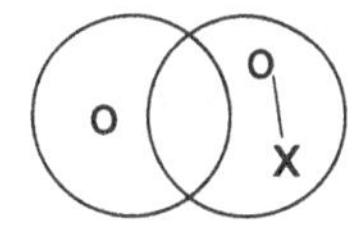

Of course, these diagrams are different from each other in that they have different signs. This shows that Peirce seems to mean semantic (*not* syntactic) equivalence among diagrams. According to him, since all of these diagrams are equivalent to each other, we should be able to transform one to another. This is the motivation behind clauses (i) and (ii-1). Intuitively, any rule that allows us to transform a diagram to another equivalent diagram must be a valid one. However, I think there are two kinds of problems with the way Peirce proposes this rule.

First, he did not have an accurate semantics to support his use of "equivalence" in a proper way. When he claims that the diagrams just given are equivalent to each other, this equivalence must have something to do with the content (or the fact) that a diagram represents, not with the way a diagram represents a fact. Therefore, we may have many syntactically different (i.e., different-looking) diagrams that are equivalent to one another. But, in the next subsection, I will show that Peirce does not maintain this correct view consistently.

Second, even though this assumption is not wrong, it is quite odd to have a transformation rule that spells out this assumption itself. An analogy to symbolic logic will reveal this oddity clearly. Suppose that a deductive system has the following inference rule: If two formulas, say α and β, are semantically equivalent, then we can deduce β from α, and vice versa. Let's take one more step from here. If we do not mind having this inference rule, why not the following rule? If formula α is a logical consequence of a set of formulas Σ, then we can deduce α from Σ. These two inference rules would put the cart before the horse, calling the necessity of a deductive system into question. A deductive system is concerned with only syntactic manipulations among formulas. If the system is sound and complete, then we know that a deductive system allows us to obtain formula α from set of formulas Σ if and only if α is a logical consequence of Σ. That is why the significance of deductive systems stems from their inference rules. Therefore, after consideration of what transformation rules are supposed to tell us, Peirce's (i) and (ii-1) of Rule 4 should be stated as the following:

(i′) *Suppose that the characters in the same compartment are of the same kind of sign whether attached or detached. Then, we may erase them except one writing of it. We may also insert the same kind of sign, whether attached or detached, in the same compartment.*

(ii-1′) *Suppose that two different signs are connected with one another. Then, we may erase them. We may also insert them in the same compartment.*

Now, we realize that part of (i′) is redundant, since we already have rule 1, erasure of a character (if detached), and rule 2, attachment of a sign. Part of (ii-1′) is also redundant, because rule 1 allows us to erase any existing sign.

It is not clear what clause (ii-2) aims to say. What does "constitutes an absurdity" mean? There are two ways to interpret this phrase. One is to take this phrase to mean that we should not be allowed to draw a diagram with two different kinds of signs in the same compartment. If so, this system has no way to represent a contradiction. Semantics should be able to tell us whether or not a diagram that the syntax of the system allows is a contradiction. For example, the syntax of a sentential language allows us to write contradictory, but well-formed, formulas, for example, $(A_1 \wedge \neg A_1)$. It is the semantics of this language that tells us that this formula is a contradiction.

Before introducing the transformation rules, Peirce makes the following comments about having these two signs, that is, "x" and "o," in the same area:

It seems also quite natural that to mark the same compartment independently with contradictory signs, as in Fig. 27, should be absurd, ...[34]

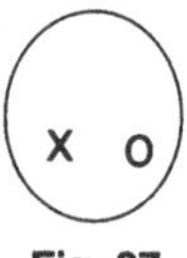

Fig. 27

His terminology, for example, *contradictory signs*, *opposite signs*, is not quite right. By contradictory signs, he might have wanted to mean signs that convey contradictory information. By opposite signs, he might have meant signs that represent opposite facts. He intended to prevent these signs (in this case, "o" and "x") from being written down in the same compartment. However, isn't it absurd to say that *to write* "$(A_1 \wedge \neg A_1)$" is absurd? What this formula aims to represent might

[34]Peirce [18], 4.359.

be absurd, but it should not be absurd to write down well-formed formulas.

The other interpretation of clause (ii-2), which seems to be more plausible, is that a diagram with "o" and "x" in the same compartment *means* absurdity. According to this interpretation, clause (ii-2) does not tell us how to transform a given diagram, but explains what assertion is made if a diagram has more than one character in a certain way. When we recall that these rules are stipulated to tell us what we are permitted to do in manipulating diagrams, it is rather puzzling why Peirce had to explain what a diagram means under these rules. What assertion is made in a diagram belongs to semantics, whereas the transformation rules belong to syntax. This clearly reveals Peirce's lack of a distinction between syntax and semantics.

Clause (iii) is the most interesting part of rule 4. This rule allows us to transform the diagram on the left to the one on the right:

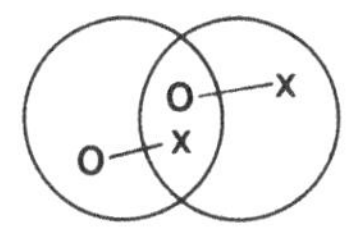

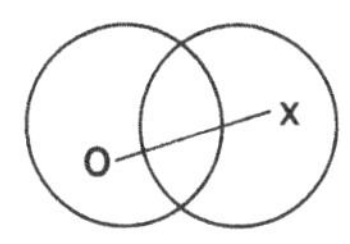

Let us see how this rule should be interpreted in order to illustrate the following syllogistic reasoning:

All S is P or some S is P.
No S is P.
Therefore, there is no S.

The first diagram in the following represents what the two premises convey.

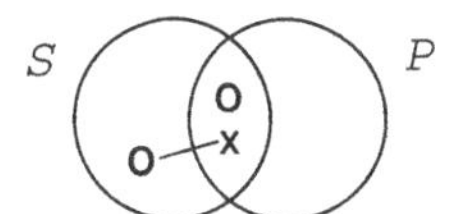

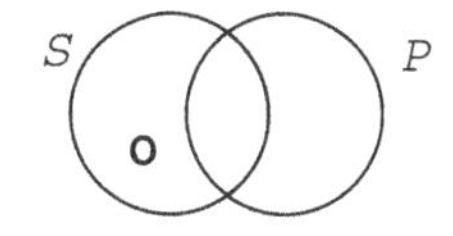

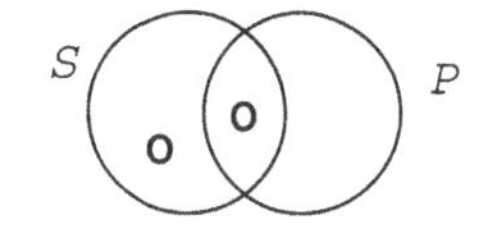

One of the o's in the first diagram is not attached to any other sign. Accordingly, the antecedent of clause (iii) is not satisfied. However, if we allow P (in clause (iii)) to be an empty sign, then we get the second diagram from the first one. After that, we need to add the second premise, "No S is P," to the the second diagram. This is how we get the third diagram, which represents the conclusion of the previous syllogism. In order to get the rightmost diagram (which we want to get), we need to represent the second premise twice.

Rule 5: We may erase a circle.

This rule is quite similar to rule 1 in that we are allowed to decrease information. Peirce specifies what is to be done in two subcases in which a circle is erased:

(i) If two compartments are thrown together containing independent zeros, those zeros may be connected.

For example, the first diagram can be transformed into the second one:

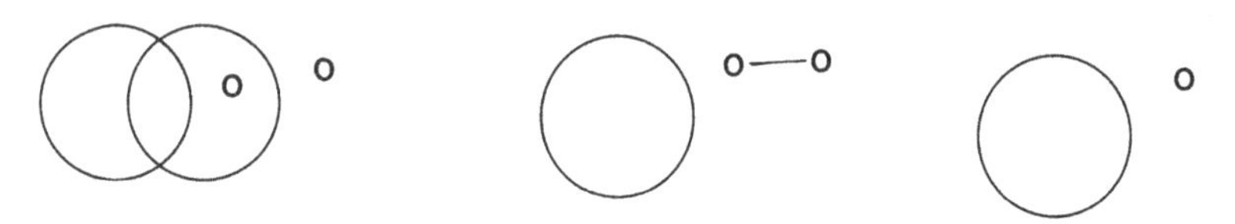

And, we apply clause (i) of rule 4 for the transformation from the second one into the third one.

(ii) If there be a zero on one side of the boundary to be erased which is thrown into a compartment containing no independent zero, the zero and its whole connexion may be erased.

Peirce gives the following transformation (the transformation of his Fig. 33 into his Fig. 34):[35]

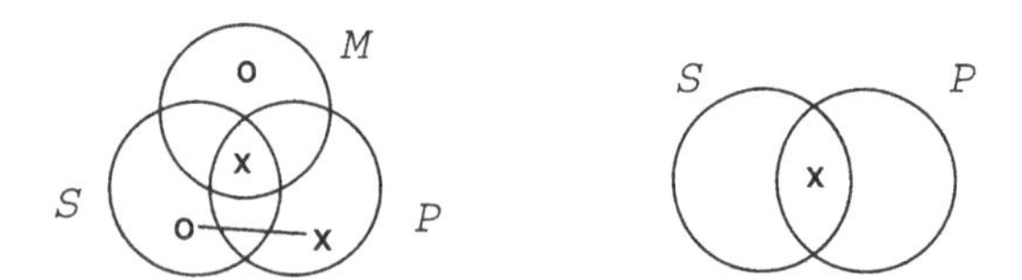

The reasoning in Example 5 can be illustrated by the following diagrams, transforming the leftmost diagram into the middle one by applying rule 5 (i) and (ii), and the middle diagram into the rightmost one by rule 4(i):

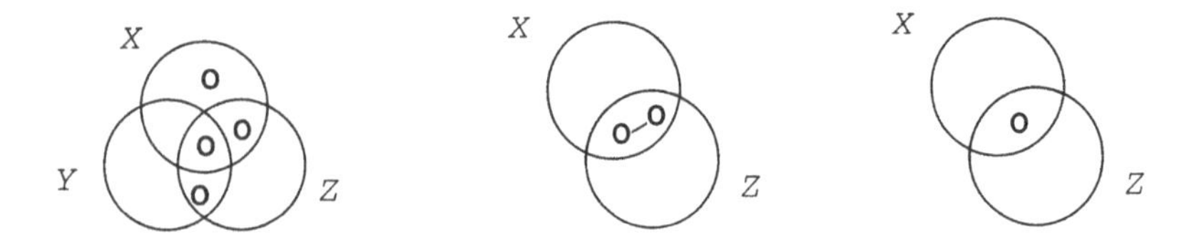

[35]Peirce [18], 4.363.

However, it is obvious that these two clauses do not exhaust the possible cases when a circle is erased. For completeness, we definitely have to specify more cases. For example, this rule does not mention x's at all. Definitely, clause (ii) would work differently in the case of "x".

Rule 6: We may draw a circle.

With this rule, the Peirce system can illustrate the valid reasoning in Example 6, whereas neither the Euler system nor the Venn system can. In the following, we get the second diagram from the first one by rule 6, the third one from the second by rule 5, and the fourth from the third by rule 1:

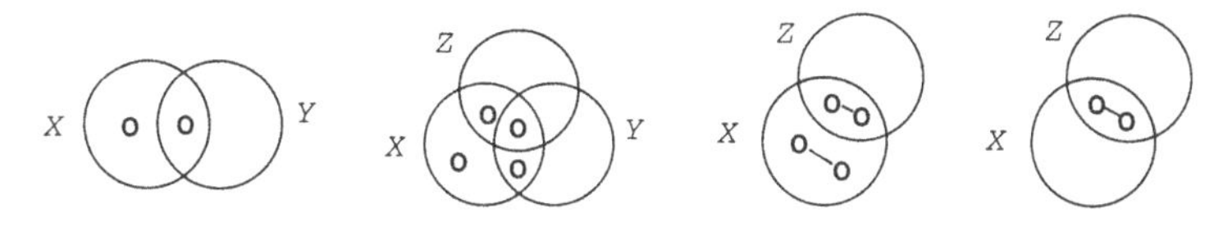

Again, in this case we need more specifications for how to change existing x's or other connected bodies.

PIR unsupported

In the previous subsection, we examined Peirce's six transformation rules. Three points should be recalled: (1) Some of the rules need to be clarified. (2) We need more rules to make this system complete. (3) Some semantic terminology (equivalence or absurdity) is used without clarification. Peirce himself was aware of the first two points. The following passage comes after his discussions of the six transformation rules:

> These six rules have been written down entirely without preconsideration; and it is probably that they might be simplified, and not unlikely that some have been overlooked.[36]

These two defects can always be improved by modifying and adding rules. The third defect is much more serious in that this reveals Peirce's lack of a distinction between syntax and semantics. In the previous subsection, I mentioned several problems that arise from this confusion. I also pointed out that Peirce relies on semantic equivalence (without any explanation) in some of his rules. In this subsection, first, I will show that this semantic use of "equivalence" is not maintained when he discusses Lambert's diagrams at length. I will also show that Peirce's approach to

[36] Peirce [18], 4.362.

diagrammatic systems has a fundamental limit, however elaborated his rules may become. Finally, I will claim that this flaw comes from his unclear notion of semantics. This diagnosis will play an important role when I present formalized Venn systems in the following chapters.

In rule 4, Peirce brings in a semantic notion, equivalence. For example, he says that a connected sign, x–o, is equivalent to no sign or that repetitions of x's are equivalent to one x. He allows transformations among these equivalent signs. That is, his semantic use of equivalence seems to justify syntactic manipulations. This shows that Peirce is quite aware of the possibility that a representation system has many different syntactic forms with the same semantic import. However, when he criticizes Lambert's linear diagrams, this possibility is absolutely forgotten.

According to Lambert's system,[37] four categorical propositions are represented as follows:

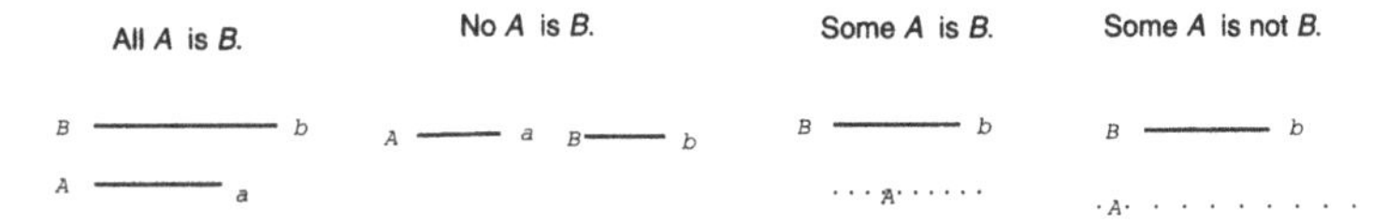

Lambert's dotted lines (for existential statements) require some explanation. Lambert represents an individual as a dot. In the case of the representations of "Some *A* is *B*" and "Some *A* is not *B*," the position of the letter "A" is important. That is, there is at least one thing that is *A* and that object is represented where the letter "A" is written. Dotted lines represent uncertainty in the following sense: Something else that is *A* might exist, and if it does, it can be represented as one of those dots.

Peirce attacks Lambert's diagrams for a particular proposition, "Some *A* is *B*."[38] Peirce claims that this diagram is incorrect and this incorrectness is just a reflection of Lambert's mistaken ideas about logical necessity and representation. According to Peirce, Lambert is one of the logicians who wrongly believe that logical necessity is a relation between the ways of thinking, not a relation between facts, and who also wrongly believe that a sentence or a diagram represents a way of thinking about a fact, not the fact itself. Therefore, Lambert would say that conclusion *C* follows from premise *P* if there is a necessary relation between the way of thinking that *C* represents and the way of thinking that *P* represents. However, Peirce claims that a logical relation should be a relation between *facts*, not a relation between *ways of thinking*, and that a sentence

[37] Lambert [15].
[38] Peirce [18], 4.353.

or a diagram represents a fact itself, not a way of thinking about the fact. Therefore, Peirce would say that conclusion C follows from premise P if there is a necessary relation between the *fact* that C represents and the *fact* that P represents.

I agree with Peirce that Lambert would be wrong if he thought that a representation system represents the ways of thinking about facts, not the facts themselves, or that logical consequence is a relation between the ways of thinking about facts. First of all, it is not clear at all how to compare the ways of thinking each of these statements represents in order to check whether or not there is a necessary relation among them. However, I do not agree with Peirce's reason why he believes that Lambert's diagram for a particular proposition, "Some A is B," is wrong. Peirce claims that Lambert's misbelief that a diagram represents a way of thinking about a fact (not the fact itself) leads him to two different diagrams in the following case:[39]

Some *A* is *B*. Some *B* is *A*.

B ——— A ———

A . . . . B . . . .

Since a diagram represents a fact, and two propositions, "Some A is B" and "Some B is A," represent the same fact, Peirce says that it is wrong to get two different diagrams in this case. In other words, Peirce thinks there should not have been a distinction between the diagram for "Some A is B" and the diagram for "Some B is A." This criticism is summarized in Peirce's following passage:

> Throughout Lambert's whole treatment of syllogistic [reasoning], the way of thinking is made the principal thing. Under these circumstances, it was impossible for him to have a clear conception of the proper nature of a system of syllogistic graphs.[40]

[39]Peirce [18], 4.353. The diagrams presented here are Peirce's misquotation from Lambert. They should have been as follows:

Some *A* is *B*. Some *B* is *A*.

B ——— A ———

. . . . A. . . . . B. . . .

[40]Peirce [18], 4.353.

Peirce might be right that the reason why Lambert got two different diagrams, as shown above, is that Lambert thought there is a difference in the way of thinking between "Some A is B" and "Some B is A." However, nothing would be wrong to get different diagrams for the representation of the same fact unless syntactically different diagrams cannot represent the same fact. Obviously, in this context, Peirce thinks that the same fact cannot be represented by different diagrams. He correctly believes that a sentence or a diagram represents a fact, but wrongly believes that the same fact should always be represented by the same diagram. Even though I agree with Peirce that a diagram represents a fact (*not* a way of thinking about a fact), I do not agree with him that the same fact need always be represented by the same diagram. The latter belief does not follow from the former one. There might be more than one diagram to represent the same fact, even when a diagram represents a fact. How would Peirce respond to the following first-order sentences?

$$\exists x P_1 x \vee \exists x P_2 x.$$

$$\exists x (P_1 x \vee P_2 x).$$

There is a difference between these two formulas, despite the fact that they represent the same fact for any given model. If we push Peirce's argument further, I am afraid we will arrive at the absurd conclusion that all logically equivalent statements should have the same syntactic form. The distinction between Lambert's two diagrams (one for "Some A is B" and the other for "Some B is A") would not matter if we had a semantics that shows us that both of them are logically equivalent and if we were permitted to transform one to the other. This is how Peirce himself justified some of his transformation rules.

It is quite surprising that Peirce allowed (syntactically) different but (semantically) equivalent diagrams in his system, but disallows these in Lambert's system. However, it is not impossible to guess how this inconsistency arises in Peirce's discussion. It is because Peirce did not have an accurate semantics for his system. This defect leads us to the following serious limit of Peirce's project on logical diagrams.

Suppose students in a logic class are asked to prove that the following simple argument is valid.

> There is no unicorn.
> Therefore, no unicorn is red.

Suppose the students gave two kinds of answers as follows:

Answer 1

1.	$\forall x \neg Ux$	Premise	
2.	$\neg Ua$	1	Universal Instantiation
3.	$\neg Ua \vee \neg Ra$	2	Disjunction Introduction
4.	Ua	Assume	
5.	$\neg Ra$	3, 4	Disjunctive Syllogism
6.	$Ua \rightarrow \neg Ra$	4, 5	Conditional Introduction
7.	$\forall x(Ux \rightarrow \neg Rx)$	6	Universal Generalization

Answer 2

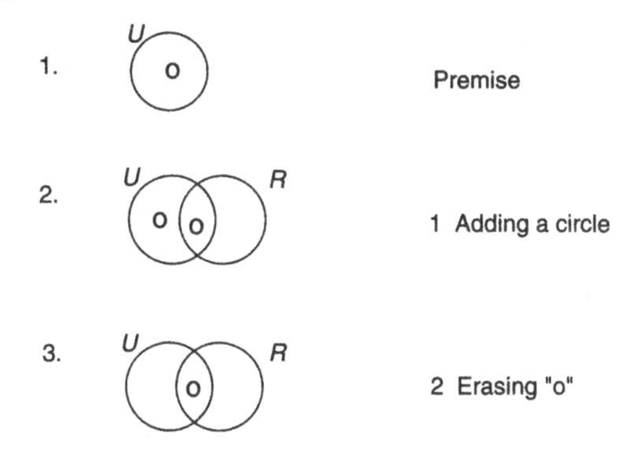

It should be noticed that each step in answer 2 uses Peirce's transformation rules. Here is an interesting question. Should we give full credit to the second answer, as we would to the first one? I think that we should *not, until* we know that answer 2 is a valid proof. Some people might object to my answer, saying that Peirce established correct rules, and the proof using only these rules is valid. This is what PIR in §2.3.2 states.[41] I think PIR is correct. However, if PIR is to be nonvacuously true, we need to know that its antecedent is true. But how does one know that these rules are correct? Peirce never tried to show the truth of this antecedent. After all the discussions about Venn diagrams, he says that "Euler diagrams [Venn diagrams] are the best aids in such cases [questions of nonrelative deductive reasoning]" (4.369). Obviously, Peirce himself realized that Venn diagrams even along with his rules of transformation cannot be used as valid proofs, but as mere "aids."

To make my point clearer, let's ask the same question about answer 1: How do we know that this answer is valid? Each inference rule is *proven*

[41] **(PIR)** If we have transformation rules that correctly guide us on how to manipulate diagrams, then we get (all and) only diagrams that are consequences of the given diagrams.

to be valid, and this assures us of the soundness of first-order logic. The semantics of a first-order language also tells us that the first sentence and the seventh sentence in this first-order language respectively represent what the premise and the conclusion in the English argument convey.

The most interesting point comes here: Nobody (including Euler, Venn, Peirce, and any other logicians who are adopting Venn diagrams in their elementary logic books) has attempted to prove that answer 2 is a valid proof just as answer 1 is, despite the fact that they have been using Venn diagrams for a long time. That is, there has been no justification for the steps taken in answer 2. Why? The answer to this question is directly related to the prejudice I addressed in the introduction. The traditional attitude toward visualization (as a heuristic tool, not as a valid proof) prevents us from making an attempt to give a proper place to visualization in the logical tradition.

Tarski's model theory justifies the validity of each rule used in answer 1. This justification convinces us that answer 1 is a real proof. Now, it seems quite obvious what we have to do in order to show that answer 2 is a *real* proof. We need to know whether Peirce's rules are valid or not. This is where Peirce stopped and is my point of departure. How do we show that this system is sound? This question requires a semantic analysis of the system. It is this requirement that Peirce was unaware of, as I have shown, and that I will carry out through two kinds of Venn systems in the following chapters.

3
Venn-I

We are about to formalize a way of using Venn diagrams. Before we present a formalism for this system (we call this system Venn-I), let us see how Venn diagrams are used to test the validity of syllogisms:

(1) Draw diagrams to *represent* the facts that the two premises of a syllogism convey. (Let us call one D_1 and the other D_2.)
(2) Draw a diagram to *represent* the fact that the conclusion of the syllogism conveys. (Let us call this diagram D.)
(3) See if we can *read off* diagram D from diagram D_1 and diagram D_2.
(4) If we can, then this syllogism is valid.
(5) If we cannot, then this syllogism is invalid.

Let us try to be more precise about each step. Step (1) and step (2) raise the following question: How is it possible for a diagram drawn on a piece of paper to represent the information that a premise or a conclusion conveys? These two steps are analogous to the translation from English to a first-order language. Suppose that we test the validity of a syllogism by using a first-order language. How does this translation take place? First of all, we need to know the syntax and the semantics of each language – English and the first-order language. We want to translate an English sentence into a first-order sentence whose meaning is the same as the meaning of the English sentence. What does it mean for two sentences to have the same meaning? The translated English sentence and the first-order sentence have the same meaning if and only if they have the same truth values in all circumstances. By this analogy with first-order languages, let us figure out how the representation takes place in step (1) and step (2). First of all, we need to know how to express ourselves in this Venn-I diagram system. After that, the semantics of this system will tell us how to get a correct representation of the information conveyed by each premise and of the information conveyed by the conclusion. Therefore, the following should be established:

(1) What are the formation rules of meaningful units in this system?
(2) About what are the meaningful units of this system?

The answer to the first question will be spelled out in the second section, and the answer to the second question will be formalized in the third section of this chapter in detail.

In step (3), what does it mean to *read off* one diagram from other diagram(s)? We read off diagram D from diagram D' if and only if D is a part of D' (or D' contains D). Accordingly, we can spell out step (4) in the following way: If we can manipulate the diagrams for the premises, that is, D_1 and D_2, to get a new diagram, say, D', so that we may read off the diagram for the conclusion, that is, D, from D', then the syllogism is valid. The next question is the following: What does it mean that we *can* or *cannot* manipulate diagrams in a certain way? These steps are comparable to the process of derivation (by using inference rules) in first-order logic. Accordingly, this question strongly suggests that we should investigate the following question, if I want to claim that this Venn-I system is deductive:

(3) What are the rules for manipulating the meaningful units of this system?

In answer to this question, six rules of transformation are introduced in the fifth section of this chapter. The set of rules of transformation allows us to manipulate some given diagrams to get a certain diagram. However, what are the desiderata for the set of rules of this system? First of all, we do not want this system to allow us to get a diagram, say, D, from a diagram, say, D', such that we cannot extract the information represented by D from the information represented by D'. That is, we want this system to be sound. This is what I will prove in the sixth section of this chapter, after presenting a formal semantics of this system. Second, we want this system to allow us to get any diagram D from diagram D' when the information represented by D is extractable from the information represented by D'. That is, we want this system to be complete. This will also be proven in the last section of this chapter.

It will be important to recall how these three issues – the formation rules, the meaning of a unit, and the transformation rules – were treated by Euler, Venn, and Peirce, respectively. Both Euler and Venn must have had their own intuitions about the second point, the meaning of a diagram. This is why both of them could start using circles for representing information. However, neither of them made a distinction between syntax and semantics. Accordingly, we cannot expect either Euler or Venn to have raised our three questions and answered them. In the previous chapter, I quoted a passage from Venn which revealed his confusion between syntax and semantics. We also know that this confusion had to wait to be cleared up for more than three decades. Peirce was almost the first person to start seeing the necessity for the third question, that is, the transformation rules. As seen in the last section of the previous chapter, Peirce himself suggested six transformation rules. He must have had an idea about the desiderata for these rules: These rules should allow all and only legitimate manipulations. However, Peirce *could not prove* that his rules do accomplish these desiderata. In order to prove this, we need to answer the second question first, that is, the semantics of this system.

Keeping the importance of all these three questions in mind, I will answer them in the following order: The first question, about the notational aspects of diagrams, is answered in §3.2, the second, about the meaning of diagrams, in §3.3, and the third, about manipulating diagrams, in §3.5. With the tools of the semantics, the legitimacy of the transformation rules will be proven in the last two sections of this chapter.

3.1 Preliminary remarks

In the next section, we will present a formal representation system which will also be proven to be sound and complete. Before precise formation rules and formal semantics are introduced, I will discuss in this section the intuitive ideas behind this representation system. This informal discussion will focus mainly on the features we want to incorporate into this system. In doing so, we will examine the following questions: What kind of diagrammatic objects are needed, how are these objects formed as meaningful units, and what are these objects to represent? While these matters are being discussed, an interesting comparison with a first-order language will be dealt with.

The following are examples of the information that this system aims to convey:

> All unicorns are red.
> No unicorns are red.

Some unicorns are red.
Some unicorns are not red.

These four pieces of information have something in common. That is, all of these are about some relation between the following sets: the set of unicorns and the set of red things. Each piece of information shows a different relation between these two sets. Therefore, the Venn-I diagram system needs to represent sets and relations between sets.

A set is represented by a differentiable closed curve that does not self-intersect, as follows:

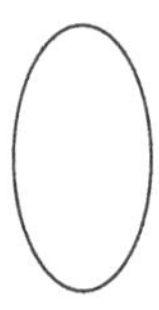

A main question is whether we want to have an infinite number of different closed-curve types *or* only one closed curve type in order to represent an infinite number of sets.

Before going through each alternative, an analogy with sentential logic might be helpful in this matter. In sentential logic we are given an infinite sequence of sentence symbols, A_1, A_2, ... An atomic sentence of English is translated into a sentence symbol. When we translate different atomic sentences of English into the language of sentential logic, we choose different sentence symbols. It does not matter which sentence symbols we use, as long as we use different symbol types for different English sentence types. Another important point is that after choosing a sentence symbol (type), say, A_{17}, for a certain English sentence (type), we have to keep using this sentence symbol type, the seventeenth sentence symbol, for the translation of this English sentence type.

There are many ways to have different closed-curve types in the system. In any case, the first alternative, having different closed-curve types, seems to have one advantage over the second alternative, having one closed-curve type. We could say, just as in the language of sentential logic, that tokens of the same closed-curve type represent the same set. However, as I will point out in the next paragraph, the problems each possible alternative has are serious enough to give up the first choice and to choose the other choice, that is, to have only one closed-curve type.

Suppose that we want to have an infinite number of closed-curve types. Then, we would come up with an infinite number of different-looking closed curves as different closed curve types. Again, there are choices.

For example, we can write down an index next to each closed curve so that different closed-curve types carry different indices. Or we can draw different sizes of closed curves to make a distinction among types. Or we can assume that there are very similar or identical-looking but many infinitely different closed-curve types. Theoretically all of these three alternatives would work. However, each case has a problem. The first alternative does not coincide with the way we actually have used Venn diagrams. We use diagrams without any index. The second choice would not be practical, since we would have to be sensitive to the size of each closed-curve token. Especially when we want to have infinitely many closed-curve types, the sizes of closed curves would become enormously big. If the third way – to have different closed-curve types that do not look different from each other – is chosen for this system, we would have to accept the following counterintuitive aspect: Very similar looking (or even identical-looking) closed-curve tokens might belong to different closed-curve types.

Now, let's move to the other alternative mentioned, that is, having only one closed-curve type. In this case, we have to solve one immediate problem, that is, how to represent different sets by one closed-curve type. Of course, we want to say that in Venn-I different sets are represented by different closed curves. However, these different closed curves are tokens of the same type – the closed-curve type, since we have only one closed-curve type. In the case of sentential logic, we have an infinite number of sentence symbol types to represent an infinite number of English sentence types. Therefore, different English sentence types are represented by different sentence symbol types. In Venn-I we have only one closed-curve type to represent an infinite number of sets. A main problem is how to tell whether given closed curves represent different sets or not. Of course, we cannot rely on how these tokens look, since every token of a closed curve belongs to one and the same type. It seems obvious that we need an extra mechanism to keep straight the relations among tokens of a closed curve, unlike in the case of sentential logic. This point will be discussed in detail in the next section after the primitive objects are introduced.

Suppose that the following closed curve represents the set of unicorns:

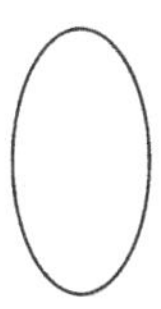

Accordingly, this closed curve makes a distinction between the set of unicorns and everything else. Strictly speaking, the area enclosed by the closed curve, not the closed curve itself, represents the set of unicorns. It will be convenient if we can treat everything else as a set as well. However, there is no such set as the set of non-unicorns, unless there is a background set. Therefore, we want to introduce a way to represent the background set in each case. Whatever sets we want to represent by closed curves, we can always come up with a background set that is large enough to include all the members of the sets represented by the drawn closed curves.[1] A background set is represented by a rectangle, as follows:

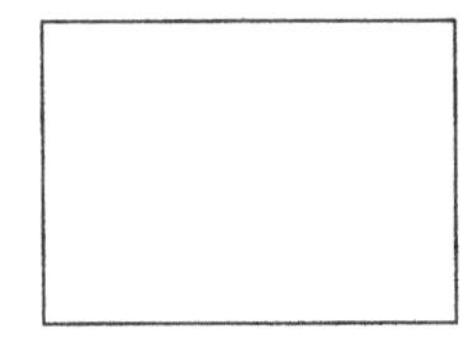

As in the case of closed curves, we also need some mechanism in order to check whether tokens of a rectangle represent the same background set or not.

For the given example, we will draw two closed curves within a rectangle to represent three sets: the set of unicorns, the set of red things, and the background set. However, in order to represent the four pieces of information mentioned earlier, that is, "All unicorns are red," "No unicorns are red," "Some unicorns are red," and "Some unicorns are not red," we should be able to represent the following sets: the set of nonred unicorns and the set of red unicorns.[2] We might need to represent the set of red nonunicorns and the set of nonred nonunicorns as well, depending upon the information we want to convey. This is the motivation behind Venn's primary diagrams, as seen in the previous chapter. The principle is to draw closed curves in such a way that we should be able to represent all of these sets in one diagram:

[1] We assume that a diagram has a finite number of closed curves.

[2] "All unicorns are red" conveys the information that the set of nonred unicorns is empty, whereas "Some unicorns are not red" conveys the opposite information, that is, the set of nonred unicorns is not empty. "No unicorn is red" says that the set of red unicorns is empty, whereas "Some unicorns are red" says the opposite.

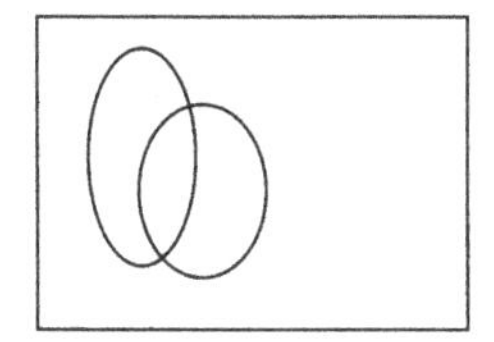

In addition to the background set, the set of unicorns, and the set of red things, the overlapping closed curves make a distinction among the set of red unicorns, the set of nonred unicorns, the set of red nonunicorns, and the set of nonred nonunicorns. This feature should be incorporated not only into the syntax of this system, that is, into the formation rules, but into the semantics of this system in the following way: Two sets represented by two disjoint areas do not share any element. And a background set, which a rectangle represents, is divided exhaustively by the sets represented by the enclosed areas that are included in the rectangle. After establishing the semantics, we will prove that these two desired features are expressed in this semantics.

So far, we have discussed how this system represents sets. Now what we need is a way to represent relations between sets. For example, the information that all unicorns are red conveys information about a certain relation between the set of unicorns and the set of red things. That is, every member of the former set is also a member of the latter set. However, this relation can be expressed in terms of the set of nonred unicorns. That is, the set of nonred unicorns is empty. In the previous paragraph, we suggested that this system should represent the set of nonred unicorns. Therefore, the problem of representing relations between sets reduces to the problem of representing the emptiness or nonemptiness of a set. For the emptiness of a set, we shade the whole area that represents the set, as Venn did. In order to represent that a set is not empty, we put down "⊗" in the area representing the set, as Peirce wrote in "x." If the set is represented by more than one area, we draw "⊗" in each area and connect the ⊗'s by lines. (As seen in Chapter 2, this idea was also suggested by Peirce.) For this, we adopt the expression $\otimes^n$ ($n \geq 1$) and call it an x-sequence. Each x-sequence consists of a finite number of ⊗'s and (possibly) lines. The formation rules of each object will be discussed in the next section in detail.

As seen in the previous chapter, Peirce adopts Venn diagrams for the representation of sets, replaces Venn's shading with the symbol "o," and adds "x" and a line to connect o's and x's. We have also seen that the connected o's in Peirce's system are so confusing that the system cannot maintain much visual power. Accordingly, Peirce's modification of Venn diagrams increases its expressive power, while sacrificing its visual power. I suggest that the Venn system that I call Venn-I and show to be sound and complete should strive for half of Peirce's modification of Venn's original system: We want to increase its expressiveness, while maintaining its visual power. Therefore, Venn-I, of which I am about to establish the syntax and the semantics, consists of the following: (i) the representation of a background set (rectangle), (ii) Venn's representation of sets and the empty set (closed curves and shading), and (iii) Peirce's representation of a nonempty set ("x" and the connected x's).

3.2 Well-formed diagrams

In this section, syntactic features of this diagrammatic system will be discussed in detail. First of all, we need to specify the set of primitive objects of which a meaningful unit in this system consists. This will be presented in the first subsection. After that, we will introduce the convention for referring to the constituents of the system in the second subsection. A characteristic of this diagrammatic system is that it will require an extra mechanism which other symbolic systems do not need. With these tools, finally, the grammar of the legitimate diagrams of this system will be formulated in the fourth subsection.

3.2.1 Primitive objects

We assume we are given the following sequence of distinct diagrammatic objects to which we give names as follows:

Diagrammatic Objects	Name
⬯	closed curve
▭	rectangle
●	shading
⊗	x
—	line

Before moving on to the formation rules of this system, we need to discuss two points about these primitive objects. One is a convention for talking *about* these diagrammatic objects. The other is the relation among tokens of a closed curve and among tokens of a rectangle.

3.2.2 Talking about the objects: pseudonaming

As we know, when we talk *about* linguistic representation systems, one difficulty is the possible confusion between the object language and the metalanguage. This confusion arises from the fact that the objects that both languages (i.e., object language and metalanguage) consist of are linguistic. To avoid this confusion, we usually adopt the convention that we name each expression by enclosing it within quotation marks. This kind of difficulty will not exist when we talk about Venn-I, since the primitive objects of this system are not linguistic, but diagrammatic.

However, a different kind of difficulty does arise. When we discuss this formal system, we should be able to mention parts of diagrams. In the case of first-order logic, as just said, we can mention any expression by the convention of naming that expression. When object language, for example,

$$\forall x(Px \rightarrow Qy),$$

is given, we can talk about a universal quantifier or a free variable by rewriting that linguistic expression and enclosing it within quotation marks – "$\forall$" or "y". Also, "$\forall x(Px \rightarrow Qy)$" is the name of the whole formula. In this visual system, we do not need any quotation marks, since the object language is diagrammatic, not linguistic. However, it will be inconvenient if we have to draw rectangles, closed curves, or shadings whenever we want to mention them. Alternatively, we might want to mention a rectangle or a closed curve in diagrams just by calling it a rectangle or a closed curve. However, this suggestion leads to another kind of problem. Suppose that there is more than one closed curve in a diagram. We cannot mention one of the closed curves in a diagram just by calling it a closed curve, since there is more than one closed curve and accordingly we would fail to refer to the closed curve we have in mind. Of course, we can mention a particular closed curve by using a description that applies only to that closed curve. However, that would not be convenient either.

Let me illustrate this point with an example. Let us temporarily assume that any finite combination of primitive diagrammatic objects of this system is a diagram. So the following is a diagram:

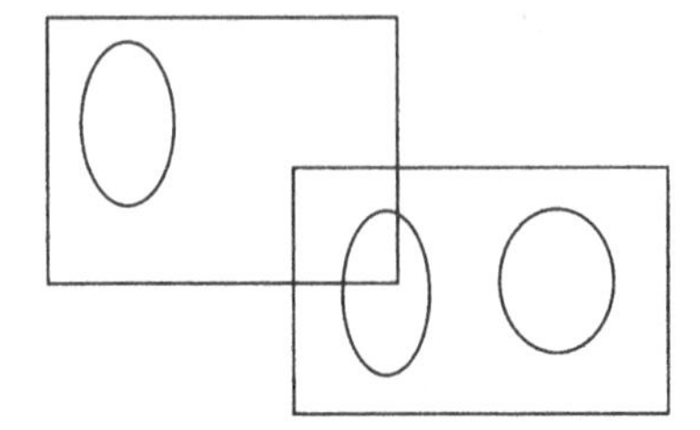

How can we talk about this diagram or parts of this diagram? For example, I want to mention the closed curve that is included only in the right rectangle. The only way to refer to that closed curve is to use a description – the closed curve that is included only in the right rectangle. Whenever we need to talk about that closed curve, we have to repeat this long description. What is worse, suppose that one more diagram which looks very similar to the previous one is given to us as follows:

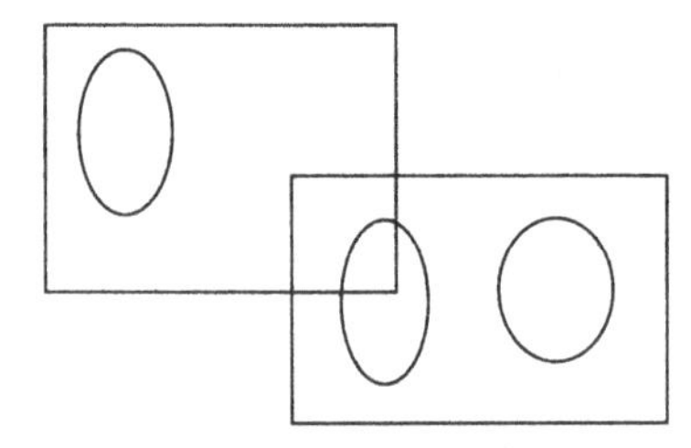

How can we refer to one of these diagrams, when it is not easy to find a description to make a distinction between these two given diagrams? Here we need a convention for naming expressions in this representation system as well.

In formulating the convention for naming the objects of Venn-I, we want to avoid the two aforementioned inconveniences – redrawing rectangles or closed curves and writing down a long description. I suggest that we write down a letter for the name of each diagram or each closed curve or each rectangle. Whenever we want to mention a diagram, a rectangle, or a closed curve, we can use the name given to the diagram, the rectangle, or the closed curve. This method helps us talk about a diagram or parts of a diagram, without using a long description.

Note that these names are not part of the language of this system. It should be emphasized that these letters as names of diagrams, rectangles, or closed curves are adopted temporarily just for our convenience. Adopting this convention, we can redraw the previous diagram this way:

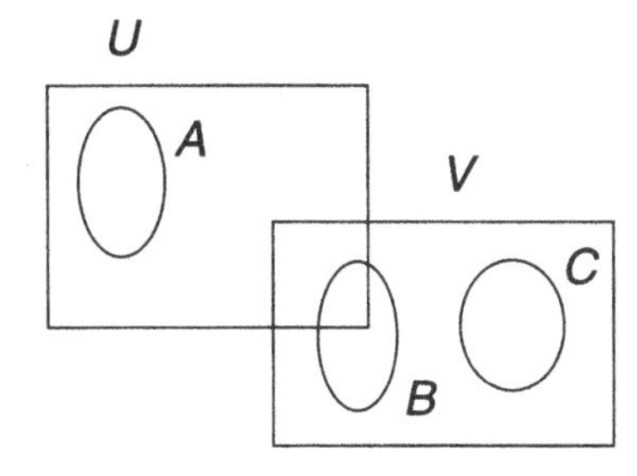

Diagram D'

Now we can talk about the closed curve that is included only in the right rectangle by its name, "closed curve C". Also, we can talk about the diagram itself by using the name of this diagram, "diagram D'," instead of redrawing the whole diagram.

Introducing letters as names (not as part of the language but as a convention for our convenience) solves our problem of how to mention diagrams, rectangles, or closed curves. But how do we mention the rest of our system, that is, shadings and x-sequences? Of course, we can introduce letters for these, as we do for diagrams, rectangles, and closed curves. However, let us try to find a way to do so without writing down names for shadings or x-sequences.

First, let me introduce three terms for our further discussion – region, basic region, and minimal region. By *region*, I mean any enclosed area in a diagram. By *basic region*, I mean a region enclosed by a rectangle or by a closed curve. By *minimal region*, I mean a region within which no other region is included. The set of regions of a diagram D (let us name it $RG(D)$) is the smallest set satisfying the following clauses:

(1) Any basic region of diagram D is in $RG(D)$.

(2) If R_1 and R_2 are in set RG(D), then the union of R_1 and R_2 is in $RG(D)$.[3]

(3) If R_1 and R_2 are in set $RG(D)$ and there is an intersection of R_1 and R_2 in D, then the intersection of R_1 and R_2 is in $RG(D)$.[4]

[3]The union of R_1 and R_2 is the region that is the set of points that belong to either R_1 or R_2.

[4]The intersection of R_1 and R_2 is the region that is the set of points that belong to both R_1 and R_2.

(4) If R_1 and R_2 are in set $RG(D)$ and there is a subtraction of R_1 from R_2 in D, then the subtraction of R_1 from R_2 is in $RG(D)$.[5]

For reasons that we will see soon, we need to refer to the regions of a diagram. We can name the regions that are made up of rectangles or closed curves by using the names of the rectangles and the closed curves. For example, in the following diagram, there are three regions – a part enclosed by a rectangle U, a part enclosed by a closed curve A, and a part enclosed by U and not by A:

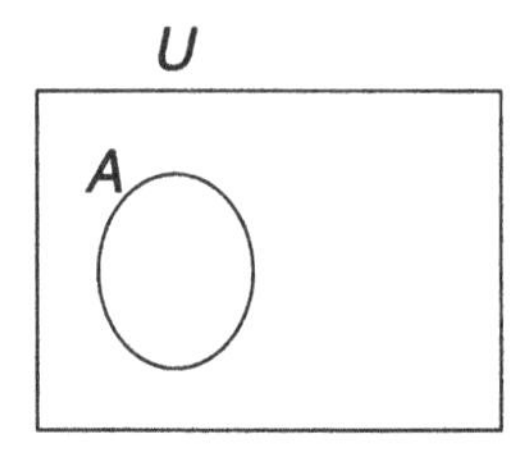

As said already, let us name a region after the name of the closed curve or the rectangle that encloses it. So, the first region is region U and the second is region A. Since the third region is the difference between region U and region A, I name it region $U - A$. Let us think of the case in which some closed curves overlap with each other, as in the following example:

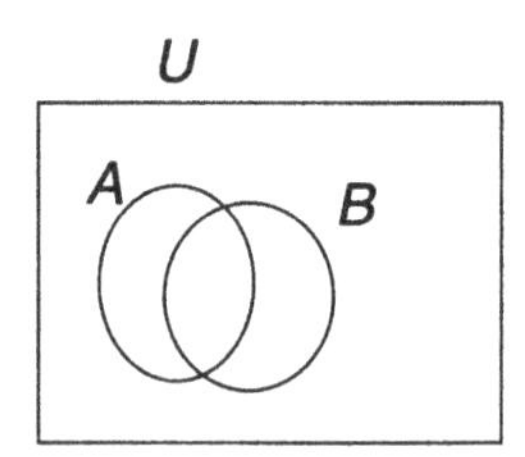

We refer to the region intersected by both region A and region B as region *A-and-B*, and refer to the region that is the union of region A and region B as region $A + B$.

To implement these ideas, we get the following convention of naming the regions:

(1) A basic region enclosed by a closed curve, say, closed curve A, or enclosed by a rectangle, say, rectangle U, is named region A or region U.

[5]The subtraction of R_1 from R_2 is the region that is the set of points that belong to R_2 but not to R_1.

(2) Let R_1 and R_2 be regions. Then, a region that is the union of R_1 and R_2 is named $R_1 + R_2$.
(3) Let R_1 and R_2 be regions. Then, a region that is the intersection of R_1 and R_2 is named R_1-*and*-R_2.
(4) Let R_1 and R_2 be regions. Then, a region that is the subtraction of R_1 from R_2 is named $R_2 - R_1$.

Recalling the definition of set $RG(D)$ (the set of regions of diagram D) given earlier, we know that this convention of naming the regions exhausts the cases.

Let us go back to the question of how to talk about shadings and x-sequences. As we will see soon, any shading or any x-sequence of any diagram (at least any interesting diagram) is in some region. Now, in order to mention these constituents of our language we can refer to them in terms of the names of the smallest regions in which they are located. For example, we can refer to a shading or an x-sequence that is in region A (where A is the smallest region with this constituent) as the shading in region A or the x-sequence in region A.

3.2.3 Counterpart relations

We assumed that diagrammatic objects are distinct from each other. For example, a shading is different from an x, and a closed curve is different from a rectangle, and so on. Furthermore, we assume that none of these diagrammatic objects is a finite sequence of other diagrammatic objects.

Symbolic logic also makes two similar assumptions. One is that each symbol is distinct from every other symbol and the other is that no symbol is a finite sequence of other symbols. These assumptions help us when presented with different tokens to tell whether they are of the same symbol type or not. For example, the following are given:

$$A_1,$$
$$A_1.$$

These two expressions are two tokens of the same symbol type, that is, the first sentence symbol. The following are tokens of different symbol types:

$$A_5,$$
$$A_6.$$

The first one is a token of the fifth sentence symbol and the second of the sixth sentence symbol. According to the first assumption mentioned (the assumption that all the primitive symbols are distinct from each other), we know that the fifth sentence symbol is different from the sixth one. Therefore, we say that these two expressions are tokens of different symbols. The other assumption, that no symbol is a finite sequence of other symbols, tells us that the following two tokens are also of different sequences of symbols:

$$A_5,$$
$$(A_6.$$

By the assumption, "A_5" cannot be elliptical for the sequence "$(A_6$." Rather we know they are different sequences of symbols.

However, in the case of Venn-I, this kind of type-token distinction is not fine grained enough. Suppose the following are given:

According to the assumption that all objects are distinct from each other, these are tokens of different types of diagrammatic objects. If we have a semantics that assigns meanings to these two different types of objects respectively, then there will be no ambiguity between a rectangle and an x-sequence. But, suppose that the following are given to us:

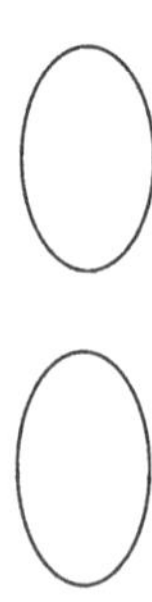

These are tokens of the same type – the closed-curve type. However, as discussed in the preliminary remarks, these two tokens might represent different sets, unlike sentence-symbol tokens of the same sentence-symbol type in sentential logic. How can we tell whether these two closed curves represent different sets or not? It depends on whether a user of the Venn diagrams intends to represent the same set or different sets by these two tokens of the closed curve. This means that the semantics that assigns meaning to a closed-curve type would not be of service to us, but the semantics that assigns a meaning to a closed-curve token would be. Accordingly, we need to specify a relation among closed-curve tokens and a relation among rectangle tokens.

The relation in which we are interested is the relation that holds among closed-curve tokens or among rectangle tokens by which a user aims to represent the same set. Let us name this relation a counterpart relation. Then, we can think of the following features for this special relation: First of all, this relation should be an equivalence relation among basic regions of given diagrams. Second, a counterpart relation holds only among tokens of the same type – among closed-curve tokens or among rectangle tokens. Third, a user would not draw two closed curves for set A and set B within one diagram unless he does not realize that set A and set B are the same. Accordingly, we want to say that within one diagram a counterpart relation does not hold among distinct basic regions. Of course, two closed curves in one diagram might represent the same set, even though a user might not realize it when he starts drawing diagrams. The following definition captures these three features:

Given diagrams $D_1, \ldots, D_n$, let a counterpart relation (let us call it set *cp*) be an equivalence relation on the set of basic regions of $D_1, \ldots, D_n$, satisfying the following:

(1) If $\langle A, B \rangle \in cp$, then both A and B are either closed curves or rectangles.

(2) If $\langle A, B \rangle \in cp$, then *either* A is identical to B *or* A and B are in different diagrams.

Within one diagram, every basic region enclosed by a closed curve or a rectangle has only one counterpart, that is, itself. Therefore, we have only one *cp* set. However, when more than one diagram is given, there is not a unique set *cp*. For example, in the following diagrams,

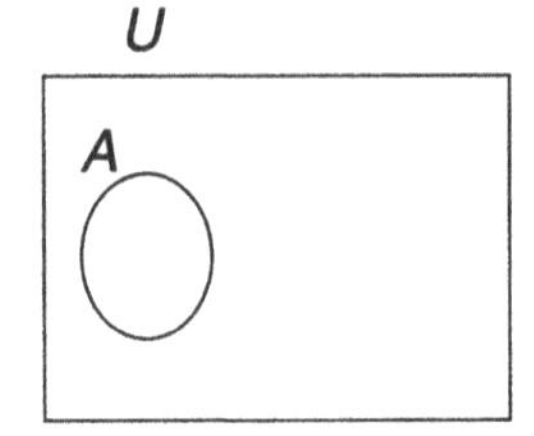

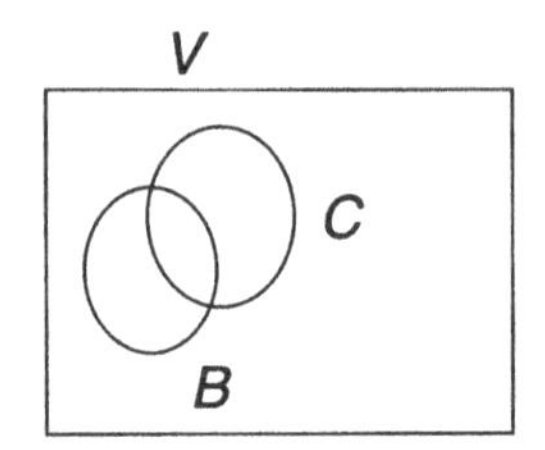

all of the following sets satisfy the defining conditions for a counterpart relation:

(1) $\{\langle U,U\rangle, \langle V,V\rangle, \langle A,A\rangle, \langle B,B\rangle, \langle C,C\rangle\}$,
(2) $\{\langle U,U\rangle, \langle V,V\rangle, \langle A,A\rangle, \langle B,B\rangle, \langle C,C\rangle, \langle U,V\rangle, \langle V,U\rangle\}$,
(3) $\{\langle U,U\rangle, \langle V,V\rangle, \langle A,A\rangle, \langle B,B\rangle, \langle C,C\rangle, \langle A,B\rangle, \langle B,A\rangle\}$,
(4) $\{\langle U,U\rangle, \langle V,V\rangle, \langle A,A\rangle, \langle B,B\rangle, \langle C,C\rangle, \langle A,C\rangle, \langle C,A\rangle\}$,
(5) $\{\langle U,U\rangle, \langle V,V\rangle, \langle A,A\rangle, \langle B,B\rangle, \langle C,C\rangle, \langle U,V\rangle, \langle V,U\rangle, \langle A,B\rangle, \langle B,A\rangle\}$,
(6) $\{\langle U,U\rangle, \langle V,V\rangle, \langle A,A\rangle, \langle B,B\rangle, \langle C,C\rangle, \langle U,V\rangle, \langle V,U\rangle, \langle A,C\rangle, \langle C,A\rangle\}$.

Among these equivalence relations, a user chooses set *cp* for each occasion. For example, if a user intends to represent the same set with A and C and the same set with U and V, then this user chooses the sixth equivalence relation as set *cp*. Some user might intend to represent the same set by A and B and the same set by U and V. In this case, the fifth relation will be the set *cp* the user chooses. Or, some user might intend to represent different sets by each closed curve and by each rectangle. In this case, the user chooses set *cp* such that all of its elements consist of a basic region and itself – the first relation given.

When more than one diagram is given to us and each diagram has at least one closed curve, only relative to set *cp* can we tell whether these closed curves represent the same set or not. The same goes for rectangles. Recall why we do not need to specify counterpart relations in sentential logic or first-order logic. The languages of these symbolic logics are defined in such a way that the tokens of the same symbol type represent the same English sentence or the same object or the same extension of a predicate. When linguistic objects are the vocabulary of a language, it is much easier to satisfy the condition that the tokens of the same type represent the same thing. However, suppose that we want to represent an infinite number of sets by some diagrammatic objects. It would not be easy to introduce an infinite number of types of diagrammatic objects.

Instead, we choose a finite number of primitive objects and introduce an extra relation among the tokens of one type. The counterpart relation just defined plays this role. Accordingly, it is crucial to know what set *cp* is when we talk about diagrams. Thus, we need to know what set *cp* is when we compare diagrams or when we want to know whether or not one diagram is obtainable from other diagrams and so on. We return to these matters in detail after defining the semantics. We also expect that the idea behind set *cp* should be implemented in the semantics of this system.

3.2.4 Well-formed diagrams

We assumed that any finite combination of diagrammatic objects is a diagram. However, not all of the diagrams are well-formed diagrams, just as not all expressions are well-formed formulas in sentential logic or first-order logic. The set of well-formed diagrams, say, $\mathscr{D}$, is the smallest set satisfying the following rules:

(1) Any unique rectangle drawn in a plane is in set $\mathscr{D}$.

(2) If D is in set $\mathscr{D}$, and if D' results by adding a closed curve interior to the rectangle of D satisfying the partial-overlapping rule (described subsequently) and the avoid-⊗ rule (described subsequently), then D' is in set $\mathscr{D}$.

Partial-overlapping rule: A new closed curve introduced into a given diagram should overlap a *proper part* of *every* existent *nonrectangle* minimal region of that diagram once and *only* once.

Avoid-⊗ rule: A new closed curve introduced into a given diagram should avoid every ⊗ of every existing x-sequence of that diagram.

(3) If D is in set $\mathscr{D}$, and if D' results by shading some entire region of D, then D' is in set $\mathscr{D}$.

(4) If D is in set $\mathscr{D}$, and if D' results by adding an ⊗ to a minimal region of D, then D' is in set $\mathscr{D}$.

(5) If D is in set $\mathscr{D}$, and if D' results by connecting existing ⊗'s by lines (where each ⊗ is in a different region), then D' is in set $\mathscr{D}$.

According to rule 1, the basic set of this inductively defined set of well-formed diagrams is a set of rectangles in planes, each of which is the only figure in its plane. Rules 2 to 5 do not introduce any additional rectangle. Accordingly, every well-formed diagram should have one and only one rectangle. Hence, the following diagrams are ruled out from the set of well-formed diagrams:

Part of rule 2, "adding a closed curve interior to the rectangle of *D*," tells us that if there is any closed curve in a diagram, it should be in the rectangle. Therefore, this definition rules out all the following diagrams as ill formed:

Let me illustrate through examples how the partial-overlapping rule in rule 2 works. By the first rule, the following is well formed:

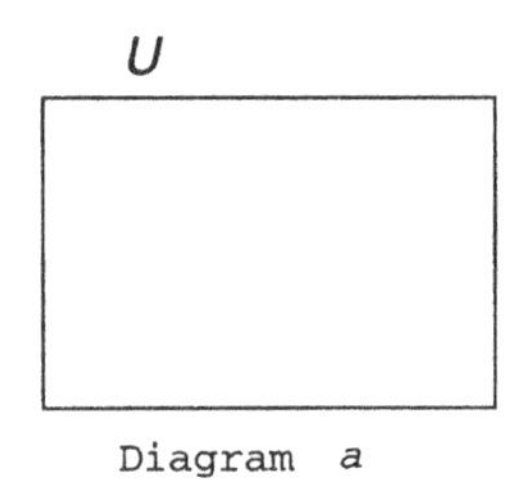

Diagram *a*

When a new closed curve, say, *A*, is introduced into Diagram *a*, rule 2 requires that three restrictions be met. It should be drawn (i) inside the rectangle, observing (ii) the partial-overlapping rule and (iii) the avoid-⊗ rule. The partial-overlapping rule applies to nonrectangle minimal regions, and the avoid-⊗ rule applies to x-sequences. However, in Diagram *a*, there is no nonrectangle minimal region and no x-sequence. Therefore, only the first condition, that is, being drawn inside the rectangle, should matter:

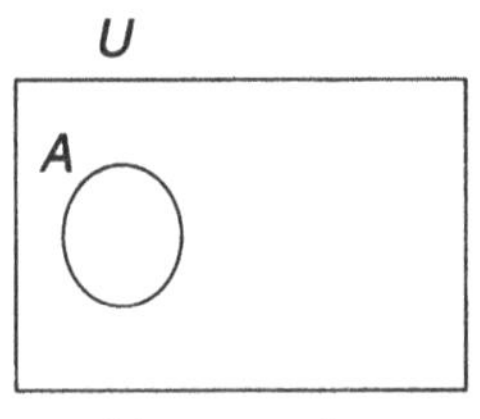

Diagram *b*

Next, I am going to draw another new closed curve, *B*, on this sheet of paper. In Diagram *b*, unlike Diagram *a*, there are two nonrectangle minimal regions: region *A* and region $U - A$. Hence, the partial-overlapping rule should be observed. According to the partial-overlapping rule, the new closed curve *B* should overlap each of these two regions, but only partially and only once. That is,

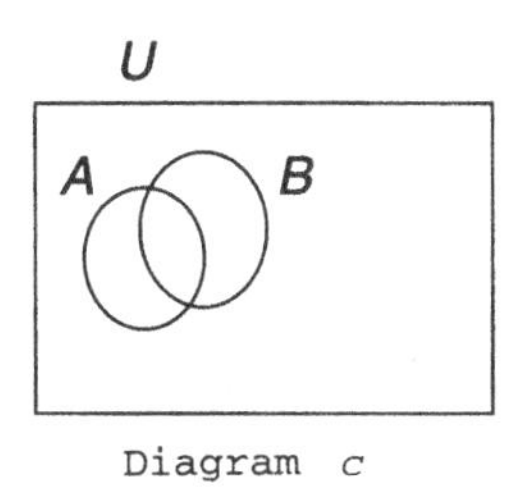

Diagram *c*

Therefore, the following diagrams are ruled out:

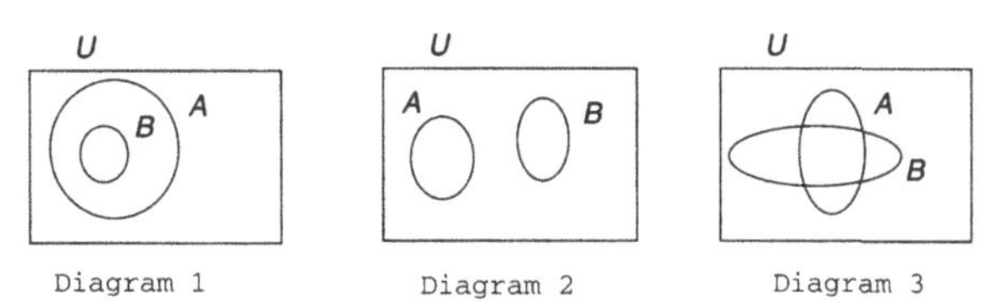

In Diagram 1, one minimal region – region $U - A$ – is not overlapped by the new closed curve *B*. In Diagram 2, one minimal region – region *A* – is not overlapped by *B* at all. In Diagram 3, minimal region $U - A$ is overlapped by the new closed curve *B* twice.

Next, we want to draw one more closed curve, *C*. All the following diagrams are eliminated:

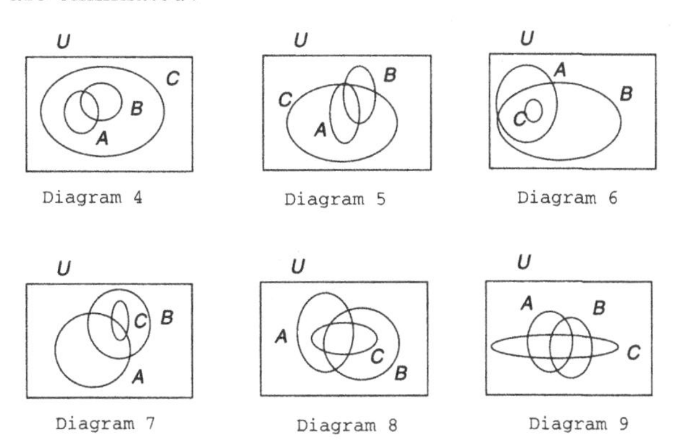

Let me go through why the partial-overlapping rule is violated in each diagram. In Diagram 4, the new closed curve C overlaps the entire area of three minimal regions – region $A-B$, region $B-A$ and region *A-and-B* – which violates the partial-overlapping rule. Similarly, in Diagram 5, the new closed curve C overlaps the entire area of two minimal regions – region $A-B$ and region *A-and-B*. There is something in common for Diagrams 6 to 8: One existent minimal region, that is, region $U-(A+B)$, is not overlapped by C. In Diagram 6, C does not overlap any part of region $A-B$ or of region $B-A$. The closed curve C and a part of region $A-B$ do not overlap each other in Diagram 7. In Diagram 9, C overlaps $U-(A+B)$ twice.

The earlier diagram *c* has four minimal regions – region $U-(A+B)$, region $A-B$, region *A-and-B*, and region $B-A$. We have to make sure that the third closed curve, C, should overlap every part of these four minimal regions. That is,

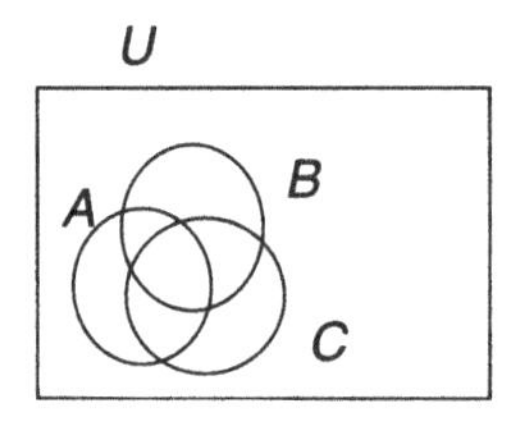

Diagram *d*

We cannot draw more than three circles satisfying this partial-overlapping condition. This is why a closed curve, rather than a circle, was introduced as a primitive object of this system. Venn himself presented a diagram with four closed curves to show that it is not impossible to draw "Eulerian circles" (differentiable, non-self-intersecting closed curves) to represent more than three terms.[6] Some have thought that it is impossible to draw more than four closed curves overlapping this way without disconnecting closed curves. This has been pointed out as a crucial shortcoming of Venn diagrams. However, V. Polythress and H. Sun proved that we can draw any finite number of convex connected curves observing the partial-overlapping rule.[7]

Now, let me show how the third condition of rule 2, that is, the avoid-⊗ rule, works. The avoid-⊗ rule requires that a new closed curve should

[6] Venn [27], pp. 127–9.
[7] Polythress and Sun [19].

not intersect any ⊗ of an existing x-sequence. Suppose we want to add a closed curve to the following diagram:

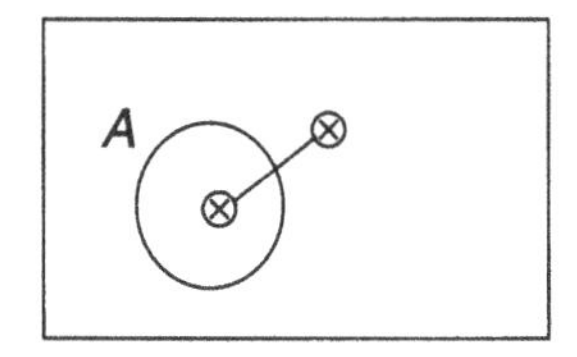

All of the following are examples where the avoid-⊗ rule is correctly applied:

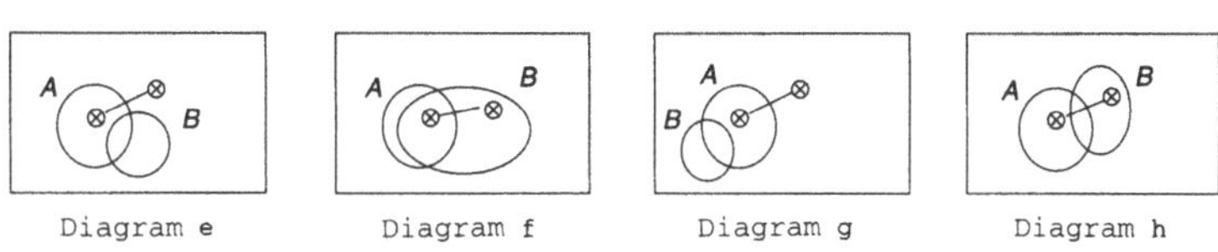

Diagram e Diagram f Diagram g Diagram h

In the first three diagrams, Diagrams *e*, *f*, and *g*, no part of the existing x-sequence is intersected by the new closed curve, *B*. In Diagram *h*, the line between two ⊗'s is intersected by *B*, but *B* does not intersect any ⊗. Therefore, it is also well formed. However, all of the following violate the avoid-⊗ rule:

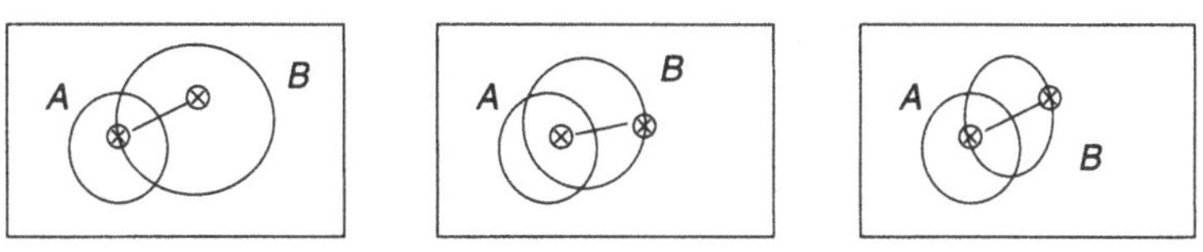

Rule 3 says that a shading, if there is any, should fill up a region(s). Therefore, the following is not well formed:

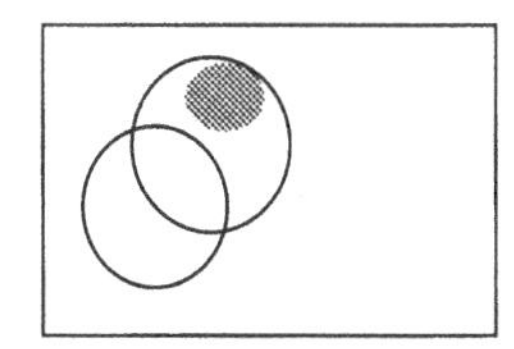

Rule 4 tells us that the following diagram cannot be well formed, since an ⊗ is not in a minimal region. It is in region *B*. However, region *B* is not a minimal region:

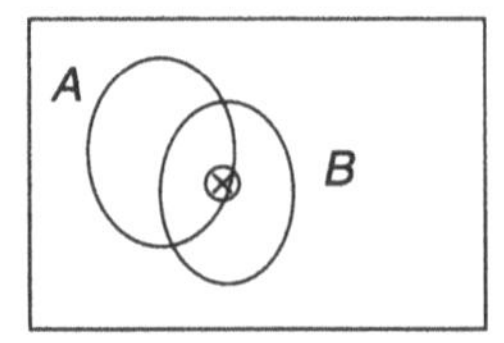

This rule says that every ⊗ should be in a minimal region. However, a minimal region may have more than one ⊗. Accordingly, the following are well formed:

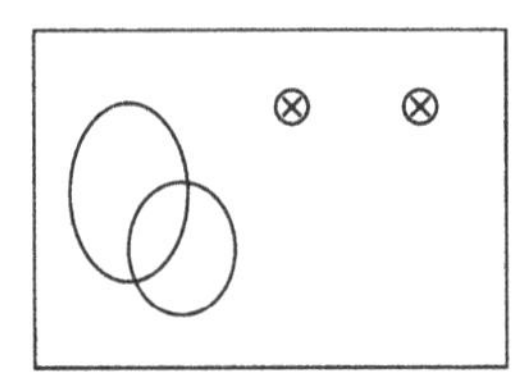

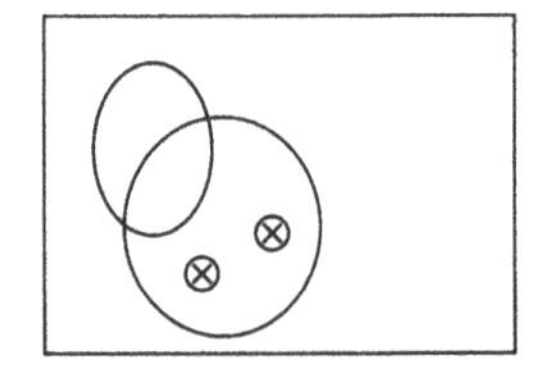

Rule 5 tells us how to construct an x-sequence, $\otimes^n$, if $n \geq 2$. Each ⊗ of an x-sequence should be in a different minimal region. Therefore, the following cannot be well formed:

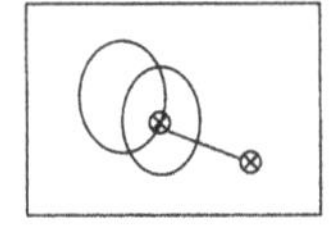

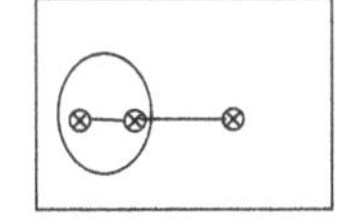

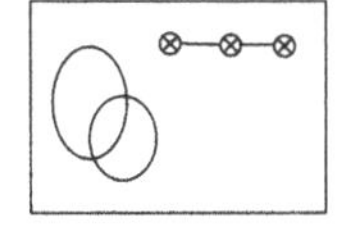

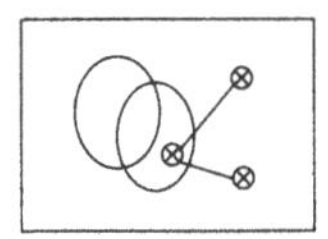

From now on, let us abbreviate "well-formed diagram" by "*wfd.*"

3.3 Semantics

A representation system is intended to represent information in terms of its own medium. When the medium is formal, we can talk about the precise formation rules of the language. That is, we can come up with a precise syntactic theory of the system. The previous section taught us how to recognize *wfd*s of representation system Venn-I. Still we cannot attach any sense to *wfd*s. Since the purpose of this system is representation, we should be able to say what this system is about. Also, we should be able to define what it means for one diagram to follow from other diagrams. These are the goals of a semantic theory of the system.

In the case of sentential logic, each sentence symbol stands for an English sentence that is either true or false.[8] Truth assignments tell us

[8]In this interpretation, the language of sentential logic seems to represent a natural language, English. Since English is another representation system, we can say that the

for what a *wff* in the sentential logic system stands. Depending upon the truth assignment chosen, sentence variables get different truth values. Since the set of *wffs* is defined inductively, each truth assignment may be extended in such a way that for a given truth assignment for sentence variables, we can calculate the truth value of a *wff* of this language. A truth assignment, say, v, satisfies a formula, say, φ, if and only if the truth value of φ is true by the extension function of v. A logical consequence between a set of *wffs* and a *wff* is defined in terms of this satisfaction relation. The formula φ is a logical consequence of the set of formulas Γ if and only if every truth assignment that satisfies every member of Γ also satisfies φ.

In the case of first-order logic, models play an analogous role. We can tell whether or not a *wff* of a first-order language is true in a given model and a given variable assignment. Predicates of a first-order language stand for different relations depending upon which model is chosen. The universal quantifier will also be interpreted as a different set depending upon the domain of the chosen model. However, sentence connectives either in sentential logic or in first-order logic get the same interpretation independently of truth assignments or models. We call them logical constants.

The situation is similar in this Venn system. As discussed before in the preliminary remarks, this system aims to represent sets and certain relations among those sets. We intend to represent sets by means of regions, the emptiness of a set by means of a shading, and the nonemptiness of a set by means of an x-sequence. A diagram might mean different things depending upon how sets are assigned to the regions of the diagram. Suppose that a set-assignment function, say, s, assigns a set to a region. Depending upon which set assignment we choose, regions represent different sets. However, a shading or an x-sequence in a region, say, A, represents the same fact about the set that is assigned to region A, regardless of set assignments. If some region is shaded in a diagram, we know that the user of the diagram meant to represent that the set that is assigned to that shaded region is empty. If some region of a diagram has an x-sequence, the user meant to represent the fact that the set assigned

language of sentential logic represents what English represents – information about the situation with which we are concerned. However, we can talk about the representational relation between the sentential-logic language and information. That is, each sentence variable in this language stands for a proposition that is either true or false. In this respect, it might be more proper to speak of propositional logic instead of sentential logic. On the other hand, as we know, there is a controversy as to whether "true" and "false" predicates apply to sentences or propositions.

to the region by a set assignment is not empty. We plan to treat these diagrammatic objects (i.e., shadings and x-sequences) as the analogues of the logical constants or predicates with fixed interpretations in symbolic logic. In this sense, they are quite similar to the identity sign in first-order logic, =.

In the first subsection, I will define a set assignment *s* that assigns sets to basic regions of diagrams, and extend this function so that we know which set is represented by a region of a diagram. (This extended function will be called $\bar{s}$.) In the second subsection, after giving the fixed interpretations of the two diagrammatic objects, that is, shading and x-sequence, I will define a satisfaction relation between a set assignment and a diagram. Finally, the consequence relation among diagrams, which is the main goal of semantics, will be defined in the third subsection.

3.3.1 Set assignments

As demonstrated in the preliminary remarks, we draw a rectangle and closed curves in order to represent some sets. We also discussed one of the advantages that Venn's primary diagrams have over Euler circles: All possible relations among sets (which are represented by the closed curves of a diagram) are representable in one diagram. This intuitive advantage should be implemented in our formal semantics. That is, we will assume that there is a set assignment *s* that assigns a set to each basic region of a diagram. Let us recall that the set of regions is defined inductively: It is closed under the operations of union, intersection, and subtraction, and the set of basic regions is the basic set.[9] Accordingly, we seem to be able to extend this assignment *s* to make a mapping from a region to a set.

The first thing to do is to formalize set assignments for basic regions of diagrams. Recalling our discussions in the preliminary remarks about background sets and counterpart relations among tokens of basic regions, we need to implement the following two constraints for a set assignment *s*:

(i) For a given domain *U*, *U* should be assigned to the rectangle of a diagram.

(ii) For a given *cp* relation, if two basic regions are *cp*-related, then the same set must be assigned to these two regions.

[9] Refer to §3.2.2.

Since set assignments should assign sets to the basic regions of well-formed diagrams, we need to define the set of all basic regions of well-formed diagrams (call it the set BRG) as follows:

Let $\mathscr{D}$ be a set of well-formed diagrams. Then,

$$BRG = \{\text{a basic region of } D \mid D \in \mathscr{D}\}.$$

Now, we define set assignment s as a function from the set BRG of all basic regions into the power set of the domain. That is, if U is a given domain and cp is a given counterpart relation, then

s: $BRG \rightarrow \mathscr{P}(U)$, where

(i) if R is a basic region enclosed by a rectangle, $s(R) = U$,[10]
(ii) if $\langle A, B\rangle \in cp$, then $s(A) = s(B)$.

Our goal is to extend this function so that a set is assigned to each region of a diagram. However, the three operations under which the set of regions is closed, that is, union, intersection, and subtraction, are not one-to-one relations. Moreover, the ranges of these operations are not pairwise disjoint. That is, the same region can be mapped from basic regions in more than one way. For example, in the following diagram, the shaded region may be mapped to *A-and-B*, *A-and-B-and-U*, or $A - (U - B)$:

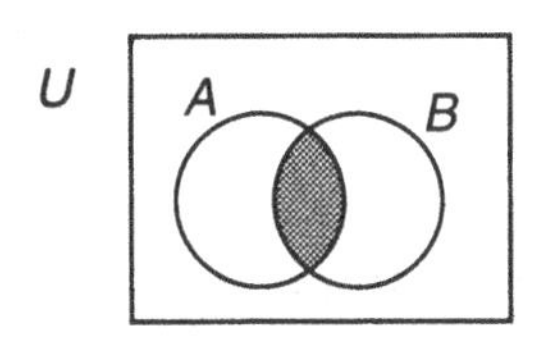

Accordingly, the extension process cannot be implemented directly at this moment. We need a detour at this stage, to prove that the result is that we have only one extension of a given set assignment s. Using the definition of s, let us define function s' such that s' assigns sets to minimal regions in the following way:

Let MRG be the set of minimal regions of *wfd*s and U be a given domain.
s': $MRG \longrightarrow \mathscr{P}(U)$, where

$$s'(A_{\text{mr}}) = (s(A_1^+) \cap \cdots \cap s(A_i^+)) - (s(A_j^-) \cup \cdots \cup s(A_k^-)),$$

where $A_1^+, \ldots, A_i^+, A_j^-, \ldots, A_k^-$ are the basic regions of a diagram such

[10]We assume that all the basic regions enclosed by the rectangles of *wfd*s are *cp*-related to each other.

that $A_1^+, \ldots, A_i^+$ are the basic regions of which A_{mr} is part, and $A_j^-, \ldots, A_k^-$ are the basic regions of which A_{mr} is not part.

We extend this function, that is, s', to assign sets to regions and name it $\bar{s}$. First, let me define RG to be the set of all regions of diagrams in set $\mathscr{D}$, where $\mathscr{D}$ is a set of well-formed diagrams. That is,

$$RG = \{\text{a region of } D \mid D \in \mathscr{D}\}.$$

Then, the function $\bar{s}$ will be defined as follows:

$\bar{s}$: $RG \longrightarrow \mathscr{P}(U)$, where

$$\bar{s}(A) = \bigcup\{s'(A_{\mathrm{mr}}) \mid A_{\mathrm{mr}} \text{ is a minimal region that is part of } A\}.$$

Now we want to check whether this function $\bar{s}$ is the extended function such that it assigns a set to a region recursively from the original set assignment s. This is, I believe, what Venn originally had in mind when he came up with the idea of primary diagrams. This is also what we assume when we draw Venn circles. Now, we want to make sure that our formal semantics will conform to this intuition. This will be done by proving that the following theorem is true:

Theorem 1 *This extended function $\bar{s}$ has the following properties:*

(1) $\bar{s}(A) = s(A)$ *if* $A \in BRG$.
(2) $\bar{s}(A) = \bar{s}(A_1) - \bar{s}(A_2)$ *if* $A = A_1 - A_2$.
(3) $\bar{s}(A) = \bar{s}(A_1) \cap \bar{s}(A_2)$ *if* $A = A_1\text{-and-}A_2$.
(4) $\bar{s}(A) = \bar{s}(A_1) \cup \bar{s}(A_2)$ *if* $A = A_1 + A_2$.

Proof Refer to the appendix.

We defined a set assignment s in such a way that for a given *cp* relation, the *cp*-related basic regions are assigned the same set. What we want is that the extended function $\bar{s}$ assigns the same set to the regions if they are *cp*-related to each other. First, we need to define *cp* relations among regions. In §3.2.3, we defined counterpart relations on the set of basic regions of *wfd*s. Now, we want to define the extended relation of *cp*, that is, $\overline{cp}$, on the set of regions of *wfd*s. Given a set *cp*, set $\overline{cp}$ is a binary relation on RG such that $\overline{cp}$ is the smallest set satisfying the following:

(1) If $\langle A, B\rangle \in cp$, then $\langle A, B\rangle \in \overline{cp}$.

Suppose that $\langle A, B\rangle \in \overline{cp}$ and $\langle C, D\rangle \in \overline{cp}$.

(2) If $A + C \in RG$ and $B + D \in RG$,[11] then $\langle A + C, B + D\rangle \in \overline{cp}$, $\langle A$-*and*-C, B-*and*-$D\rangle \in \overline{cp}$, $\langle A - C, B - D\rangle \in \overline{cp}$, and $\langle C - A, D - B\rangle \in \overline{cp}$.

The following property about $\bar{s}$ will be proven easily by the induction of set $\overline{cp}$:

Theorem 2 *If* $\langle A, B\rangle \in \overline{cp}$, *then* $\bar{s}(A) = \bar{s}(B)$.

The following feature about this set assignment will also be used for further discussion:

Theorem 3 *If region R_1 is a part of region R_2, then the set region R_1 represents is a subset of the set region R_2 represents.*

Proof Since R_1-*and*-$R_2 = R_1$, $\bar{s}(R_1$-*and*-$R_2) = \bar{s}(R_1)$. Accordingly, $\bar{s}(R_1$-*and*-$R_2) = \bar{s}(R_1) \cap \bar{s}(R_2) = \bar{s}(R_1)$. Therefore, $\bar{s}(R_1) \subseteq \bar{s}(R_2)$. □

In the preliminary remarks, we wanted two features to be incorporated into the semantics of this system: One is that any two nonoverlapping regions of a *wfd* represent disjoint sets. The other is that given a background set, the sets represented by the minimal regions of a diagram exhaust the background set. Let us check whether the set assignment just defined can take into account these two features.

Theorem 4 *For any regions R_1 and R_2 of a wfd, if R_1 and R_2 do not overlap each other, then the set represented by R_1 and the set represented by R_2 are disjoint.*

Proof Refer to the appendix.

[11]Why do we need this condition? Suppose the following two *wfd*s are given with the following *cp* relation:
$cp = \{\langle A, B\rangle, \langle A, A\rangle, \langle B, B\rangle, \langle B, A\rangle\}$.

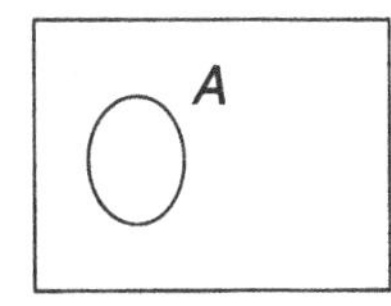

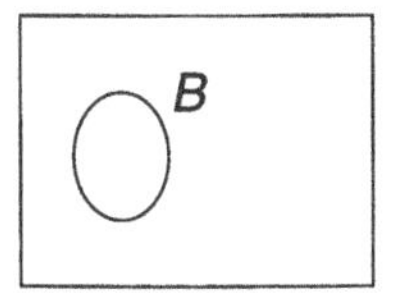

Without the condition that $A + B \in RG$, we would have to say that $\langle A + A, A + B\rangle \in \overline{cp}$, which we do not want. However, since $\langle A, B\rangle \in cp$ and $A \neq B$, A and B must be in different diagrams by the definition of counterparts. Therefore $A + B \notin RG$.

Since any two distinct minimal regions do not overlap each other, this theorem yields the following corollary:

Corollary 4 *Every pair of distinct minimal regions represent disjoint sets.*

Theorem 5 *For every region R of a wfd, the set represented by R is the union of the sets represented by the minimal regions of which R consists (i.e.,* $\bar{s}(R) = \bigcup\{\bar{s}(R_{mr}) \mid R_{mr}$ *is a minimal region and* R_{mr} *is a part of* $R\}$*).*

Proof Refer to the appendix.

By our definition of well-formed diagrams, every *wfd* has one and only one rectangle and this rectangle consists of all minimal regions of the *wfd*. If we apply Theorem 5 to the basic region enclosed by the rectangle of a *wfd*, then we get the following property:

Corollary 5 *Let U be the basic region made by the rectangle of a wfd D. Then,* $\bar{s}(U) = \bigcup\{\bar{s}(U_{mr}) \mid U_{mr}$ *is a minimal region of* $D\}$.

By Corollary 4 and 5, our formalism of what each region stands for seems to satisfy Venn's following original intuition:

What we ultimately have to do is to break up the entire field before us into a definite number of classes or compartments which are mutually exclusive and collectively exhaustive.[12]

3.3.2 Representing facts

Given set assignment s as defined earlier, we want to define what it is for a set assignment to satisfy some fact about a diagram. Like other representation systems, some of the facts about Venn diagrams have no representational import. For example, a difference in size of closed curves or x-sequences does not have any effect on what diagrams represent. However, which region is shaded or which region has an x-sequence is definitely important. These are two kinds of representing facts in this system that we need to formalize in this subsection. That is, we want this formal semantics to capture the following: (i) If region A is shaded in diagram D, then this diagram represents the fact that whatever set the set assignment assigns to region A, that set is empty. (ii) If region

[12]Venn [27], p. 111. As seen in this quotation, there is no clear distinction between syntax and semantics in Venn's writing. (Refer to §2.2. His *classes* can be read as sets and *compartments* can be read as regions in our terminology.)

A has an x-sequence in diagram D, then this diagram represents the fact that whatever set the set assignment assigns to region A, that set is not empty.

Here I want to make a notational suggestion. For our further purposes, we want to make a set out of representing facts of a diagram, compare different diagrams' representing facts, find conflicting pieces of information among diagrams, and so on. In order to facilitate these operations, I will introduce information-theoretic terminology – infons – and define them as follows:

Let A be a region.

$\langle\langle$Shading, A; 1$\rangle\rangle$ iff region A is entirely shaded.
$\langle\langle \otimes^n, A; 1 \rangle\rangle$ iff region A has an x-sequence $\otimes^n$.

An infon is a unit of information. Accordingly, this notation provides us with a convenient way to deal with representing facts.

Before making a set of representing facts, we need to clarify the meaning of the statement that a region has an x-sequence. The well-formedness of diagrams requires that a shading should fill up a minimal region entirely. We say that a region A is shaded if and only if a shading fills up the whole of region A. Therefore, whether a region is shaded or not is not ambiguous at all. However, whether a region has an x-sequence or not can be ambiguous. For instance, if an x-sequence is in region A, then we can say that the x-sequence is also in region B if region A is a part of region B. More concretely, it is true that the x-sequence $\otimes^2$ is in region A in the following diagram:

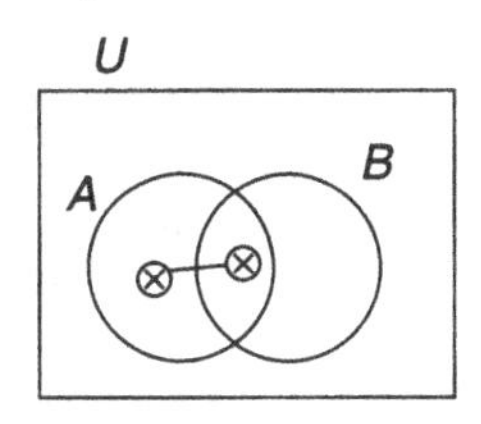

However, it is also true that the x-sequence is in region $A + B$ and in region U. In order to avoid this ambiguity, we say that an x-sequence is in region A if and only if A is the *smallest* region with the x-sequence. Since each $\otimes$ of the x-sequence is in a minimal region in a *wfd* (by the definition of well-formedness), the union of the minimal regions with $\otimes$ is the smallest region with the x-sequence.

Now, we give the definitions of the two kinds of representing facts as follows:

Let A be a region.

(1) $\langle\langle$Shading, A; 1$\rangle\rangle$ iff region A is entirely shaded.
(2) $\langle\langle \otimes^n, A; 1\rangle\rangle$ iff A is the smallest region with $\otimes^n$.
(I.e., $A = A_1 + \cdots + A_n$, where A_i $(1 \leq i \leq n)$ is a minimal region with $\otimes$ of $\otimes^n$.)

As discussed before, we can fix the interpretation of a shading or of an x-sequence across set assignments. Therefore, we define what it is for set assignment s to satisfy infon α (i.e., $s \models \alpha$) in the following way:

$$s \models \langle\langle \text{Shading}, A; 1\rangle\rangle \quad \text{iff} \quad \bar{s}(A) = \emptyset.$$
$$s \models \langle\langle \otimes^n, A; 1\rangle\rangle \quad \text{iff} \quad \bar{s}(A) \neq \emptyset.$$

What we aim to formalize in this subsection is what it is for a set assignment to satisfy a diagram. But what we mean by a diagram being satisfied by a given set assignment is that the representing facts of a diagram are satisfied by the set assignment. Since we just defined what it is for a set assignment to satisfy a representing fact, what we need to do is to make a set out of the representing facts of a given diagram. The set of the representing facts of a well-formed diagram D should be a union of all the infons such that the infon is either $\langle\langle$Shading, $x; 1\rangle\rangle$ (where x is a shaded region in D) or $\langle\langle \otimes^n, y; 1\rangle\rangle$ (where y is a region with $\otimes^n$ in D). Since which regions are shaded or which regions have x-sequences are important in this system, it will be quite useful to present a *wfd* with these two kinds of information. In addition, if we add one more kind of information, that is, which basic regions a diagram has, then this will be another way to specify a well-formed diagram without actually showing the diagram. That is, we can specify a *wfd* by concatenating the set of the basic regions, the set of the shaded regions, and the set of the smallest regions with an x-sequence. Let us call this $Seq(D)$ (we say the sequence of diagram D) and define it as follows: (Let D be a *wfd*.)

$Seq(D) = \langle BR(D), RShading(D), R\otimes^n(D)\rangle$, where

$BR(D)$ is the set of the basic regions of D,
$RShading(D)$ is the set of the shaded regions of D,
$R\otimes^n(D)$ is the set of the smallest regions[13] with an x-sequence of D.

The following is the example for this new notation:

[13] Refer to the earlier discussion about the ambiguity involving x-sequences.

Example 8 $Seq(D_1) = \langle\{U, A_1, A_2\},\ \{A_1\text{-}and\text{-}A_2\},\ \{A_1, A_2 - A_1\}\rangle$.

$Seq(D_2) = \langle\{U, A_1, A_2\},\ \{A_1\text{-}and\text{-}A_2\},\ \{(A_1 - A_2) + (A_2 - A_1)\}\rangle$.

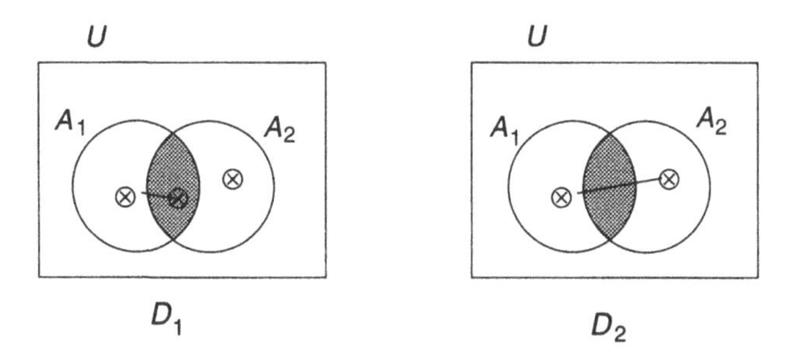

The sequence of a diagram will be used conveniently when diagrams are compared (especially) at the syntactic level in the next section. Also, we can use this notation to give the following formal definition of the set of representing facts discussed previously:

Let $Seq(D) = \langle BR(D),\ RShading(D),\ R\otimes^n(D)\rangle$. Then,

$RF(D) = \{\langle\langle \text{Shading}, x; 1\rangle\rangle \mid x \in RShading(D)\} \cup \{\langle\langle \otimes^n, y; 1\rangle\rangle \mid y \in R\otimes^n(D)\}$.

Finally, let us define the satisfaction relation between a set assignment and a diagram. We want to say that set assignment s satisfies *wfd* D if and only if s satisfies all the representing facts of D. If $RF(D)$ is a set of the representing facts of diagram D, then

$$s \models D \qquad \text{iff} \qquad \forall_{\alpha \in RF(D)}\ s \models \alpha.$$

3.3.3 Consequence relation

In this subsection, we want to formalize what it means for one *wfd* to follow from other *wfd*s. For example, our formalized semantics should allow us to say that in the following example D_2 follows from D_1:

Example 9 *Let* $\langle A_1, A_3\rangle, \langle A_2, A_4\rangle \in cp$.

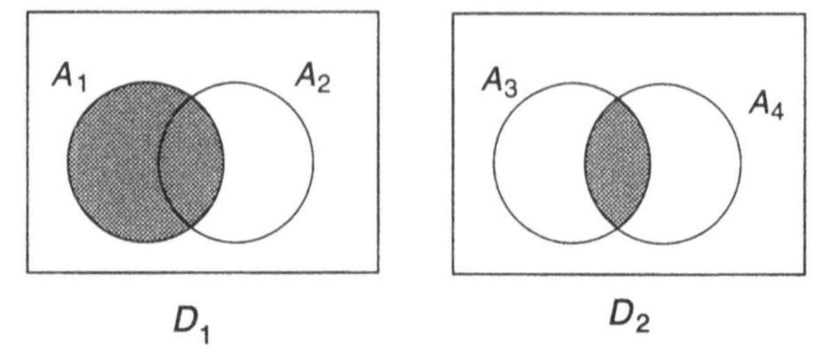

Whatever sets are assigned to the regions of these two diagrams in the preceding example, as long as cp-related basic regions get the same set (which is the case by our definition of set assignment s), what is represented by D_2 follows from what is represented by D_1.

The consequence relation among diagrams is defined as follows:

Wfd D follows from set of *wfd*s Δ (i.e., $\Delta \models D$) if and only if every set assignment that satisfies every member of Δ also satisfies D (i.e., $\forall_s(\forall_{D' \in \Delta} s \models D' \longrightarrow s \models D)$).

Let us go back to Example 9: $RF(D_1) = \{\langle\langle \text{Shading}, A_1; 1\rangle\rangle, \langle\langle \text{Shading}, A_1 - A_2; 1\rangle\rangle, \langle\langle \text{Shading}, A_1\text{-}and\text{-}A_2; 1\rangle\rangle\}$ and $RF(D_2) = \{\langle\langle \text{Shading}, A_3\text{-}and\text{-}A_4; 1\rangle\rangle\}$. Since $\langle A_1, A_3\rangle$, $\langle A_1, A_2\rangle \in cp$, $\bar{s}(A_1\text{-}and\text{-}A_2) = \bar{s}(A_3\text{-}and\text{-}A_4)$ (by Theorem 2). Accordingly, $\forall_s(s \models \langle\langle \text{Shading}, A_1\text{-}and\text{-}A_2; 1\rangle\rangle \longrightarrow s \models \langle\langle \text{Shading}, A_3\text{-}and\text{-}A_4; 1\rangle\rangle)$. Therefore, $\forall_s(s \models D_1 \longrightarrow s \models D_2)$.

3.4 Equivalent diagrams and equivalent representing facts

So far, we have formalized the following: how to recognize legitimate diagrams in this system, what a *wfd* is about, and what it means for one diagram to follow from other diagrams. As emphasized in the first chapter of this book, no attempt at a formal semantics of Venn diagrams has previously been made. Accordingly, neither the soundness nor the completeness of the system could be examined. The semantics formalized in the previous section will enable us to check whether the rules that will be presented in the next section are valid and complete.

This section will provide us with important concepts and notations which will be used conveniently for the formalization of obtainability among diagrams and the proofs of the soundness and the completeness of the system. For these following tasks, we need to compare *wfd*s in terms of their syntactic features as well as in terms of their semantic content. The first subsection will deal with syntactic comparison, and the second subsection discusses semantic comparison. More interestingly, we will observe relations between syntactic and semantic comparisons among diagrams, which will be proved in the second subsection as well.

3.4.1 Equivalent diagrams

In §3.2.3, we discussed why in Venn-I we need a more fine-grained relation than the type-token distinction for closed curves or for rectangles. This is a counterpart relation. Now let us extend this relation to the discussion on the relation among diagrams.

In the case of sentential logic or first-order logic, the token-type distinction for symbols is extended to a token-type distinction for formulas. For example, the following two *wffs* are tokens of the same formula type:

$$(A_1 \rightarrow A_2),$$
$$(A_1 \rightarrow A_2).$$

The sequence of the first expression is $\langle (, A_1, \rightarrow, A_2,) \rangle$ and the sequence of the second expression is $\langle (, A_1, \rightarrow, A_2,) \rangle$. The first element of the first sequence and the first element of the second sequence are of the same symbol type, that is, the left parenthesis. The second elements of each sequence are also of the same symbol type, that is, the first sentence symbol and so on. Therefore, we say that these two expressions are tokens of the same formula type.

Similarly, let us try to define a relation among diagrams in terms of a counterpart relation among basic regions defined in §3.2.3. Suppose the following two diagrams are given:

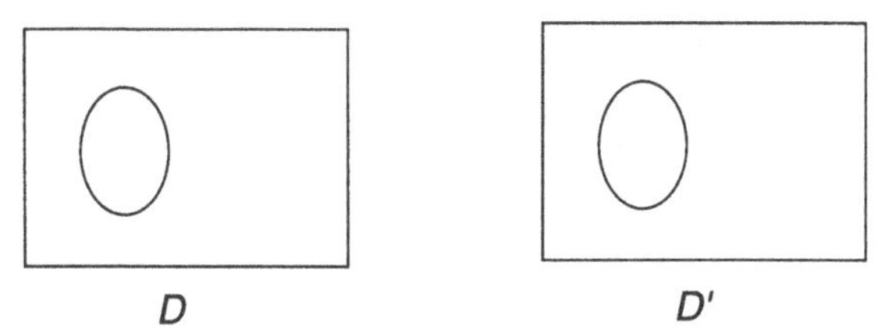

These two diagrams are distinct from each other in that they are two diagrams. In this strict sense, no diagram is the same as another. They are two different tokens. On the other hand, they consist of the same diagrammatic-object types – the rectangle type and the closed-curve type. However, we want to make a distinction between the following two cases:

(Case 1) Suppose that after having drawn diagram *D*, I intended to copy it and drew diagram D'. In this case, we want to say that some kind of identity relation holds between these two diagrams. Let us call this relation "equivalence."

(Case 2) Suppose that a user drew diagram D to represent the set of unicorns by a closed curve and drew diagram D' to represent the set of red things by another closed curve. In this case, even though they look very similar to each other, we do not want to say that an equivalence relation holds between these two diagrams.

However, this equivalence relation should be distinct from the equivalence relation in representing facts. The following examples illustrate why we need this distinction.

Example 10 *Suppose A and B represent the same set.*

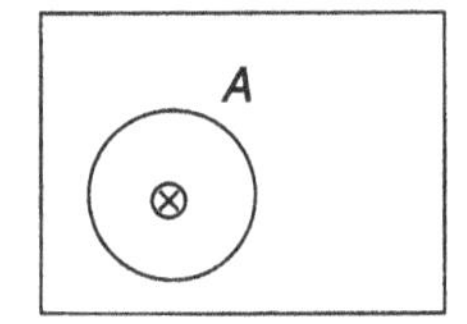

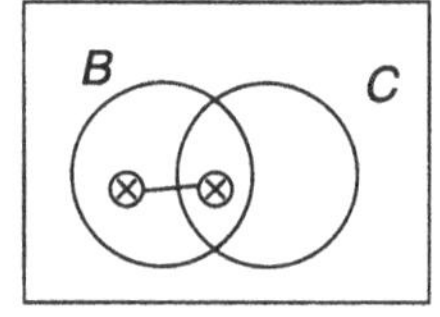

Example 11 *The following diagrams have the same set of representing facts, that is, the empty set.*

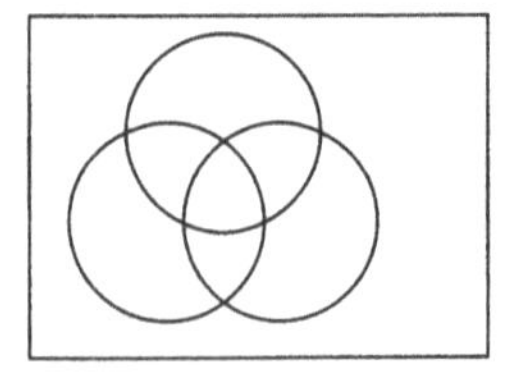

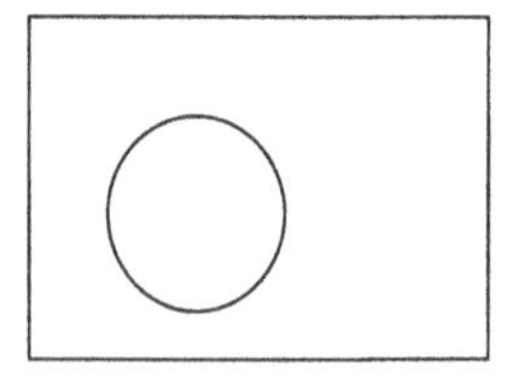

However, in both cases, we do not want to say that each pair of diagrams are (syntactically) equivalent to each other.

In this subsection, we will define this kind of a syntactic equivalence relation among diagrams. First, let us recall the case of sentential logic. Whether two expressions are tokens of the same formula type or not depends upon whether or not each element of one sequence is of the same symbol type as each corresponding element of another sequence.

In §3.3.2, we defined the sequence for a well-formed diagram, say D, as follows: $Seq(D) = \langle BR(D),\ RShading(D),\ R\otimes^n(D)\rangle$.[14]

Now we need a mechanism to compare each element of the sequences of *wfd*s. It should be noticed that each element of the sequence of a diagram is a set of regions with some property. The extended relation $\overline{cp}$, defined in §3.3.1, will be useful, since this relation tells us whether or not two regions in different diagrams are equivalent to each other. That is, if the pair of two regions is in set $\overline{cp}$, then we say that both regions are equivalent to each other or both of them are $\overline{cp}$-related to each other.

Let R_1 and R_2 be regions. Given a set cp, we define the equivalence of two regions (i.e., $R_1 \equiv_{cp} R_2$) as follows:

$$R_1 \equiv_{cp} R_2 \quad \text{iff} \quad \langle R_1, R_2\rangle \in \overline{cp}.$$

Similarly, we want to say that two sets of regions are equivalent to each other if and only if each member of one set has an equivalent region in the other set and vice versa. In addition to an equivalence relation, we will define two more relations, that is, inclusion and proper inclusion between sets of regions.

Let B and C be the sets of regions. Given a set cp, we define the inclusion relation and equivalence relation among two sets as follows:

$B \subseteq_{cp} C$ (B is *included* in C) iff $\forall_{x\in B}\exists_{y\in C}\ (\langle x, y\rangle \in \overline{cp})$.

$B \equiv_{cp} C$ (B is *equivalent* to C) iff $B \subseteq_{cp} C$ and $C \subseteq_{cp} B$.

$B \subset_{cp} C$ (B is *properly included* in$C$) iff $B \subseteq_{cp} C$ and $B \not\equiv_{cp} C$.

Now, by using these definitions, let us define the following relations between two diagrams: subdiagram, equivalence, and proper subdiagram.

Let $Seq(D_1) = \langle BR(D_1),\ RShading(D_1),\ R\otimes^n(D_1)\rangle$ and $Seq(D_2) = \langle BR(D_2),\ RShading(D_2),\ R\otimes^n(D_2)\rangle$. Then,

$D_1 \subseteq_{cp} D_2$ (D_1 is a *subdiagram* of D_2) iff $BR(D_1) \subseteq_{cp} BR(D_2)$, $RShading(D_1) \subseteq_{cp} RShading(D_2)$, and $R\otimes^n(D_1) \subseteq_{cp} R\otimes^n(D_2)$.

$D_1 \equiv_{cp} D_2$ (D_1 is *equivalent* to D_2) iff $D_1 \subseteq_{cp} D_2$ and $D_2 \subseteq_{cp} D_1$.

$D_1 \subset_{cp} D_2$ (D_1 is a *proper subdiagram* of D_2) iff $D_1 \subseteq_{cp} D_2$ and $D_1 \not\equiv_{cp} D_2$.

[14] Where $BR(D)$ is the set of the basic regions of D, $RShading(D)$ is the set of the shaded regions of D, $R\otimes^n(D)$ is the set of the smallest regions with an x-sequence of D.

This definition reflects our following intuitions about equivalent diagrams: Every *wfd* is equivalent to itself (reflexivity). If D is equivalent to D', then D' is also equivalent to D (symmetry). Given *wfd* D, if I intended to copy it and drew another *wfd* D', then according to set *cp* in my mind, D and D' are equivalent to each other. Suppose that another *wfd* D'' was drawn to copy D'. In this case, we want to say that D and D'' are equivalent to each other as well. The definition of the equivalence relation on *wfd*s tells us that this relation is transitive. Let us go through some examples to illustrate the previous definitions:

Example 12 *Let* $\langle A_1, A_2 \rangle \in cp$.

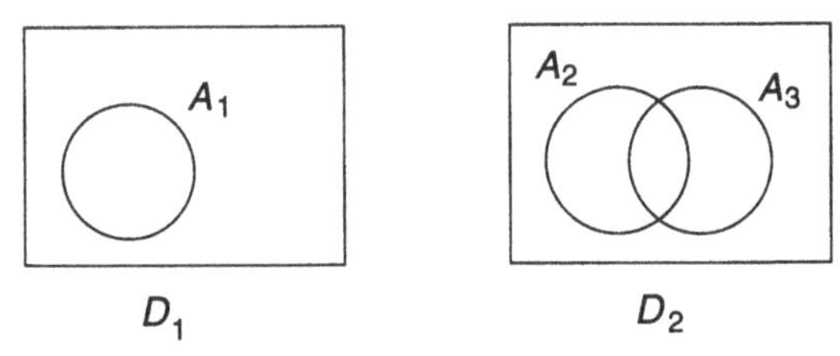

$D_1 \subset_{cp} D_2$, since $D_1 \subseteq_{cp} D_2$ but $D_1 \not\equiv_{cp} D_2$ (there is no basic region in D_1 that is equivalent to A_3).

Example 13 *Let* $\langle A_1, A_2 \rangle \in cp$.

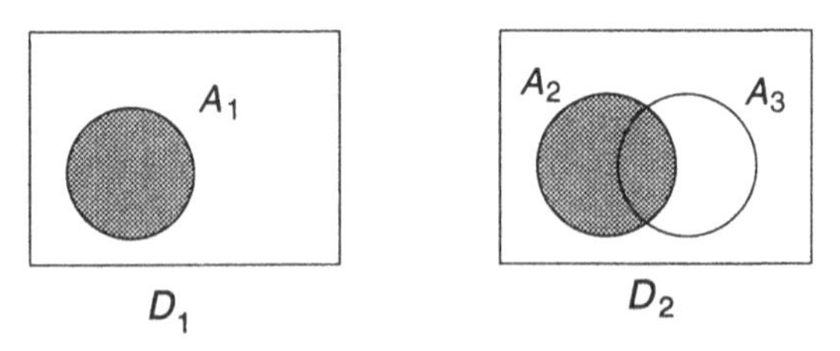

$D_1 \subset_{cp} D_2$, since $D_1 \subseteq_{cp} D_2$ but $D_1 \not\equiv_{cp} D_2$ (there is no basic region in D_1 that is equivalent to A_3 and there is no shaded region in D_1 that is equivalent to a shaded region of D_2, A_2-and-A_3).

Example 14 *Let* $\langle A_1, A_2 \rangle \in cp$.

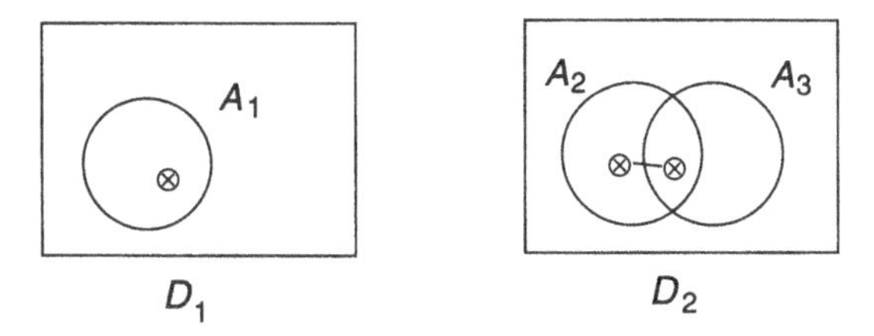

$D_1 \subset_{cp} D_2$, since $D_1 \subseteq_{cp} D_2$ but $D_1 \not\equiv_{cp} D_2$ (there is no basic region in D_1 that is equivalent to A_3).

Example 15 *Let $\langle A_1, A_3 \rangle$, $\langle A_2, A_4 \rangle \in cp$.*

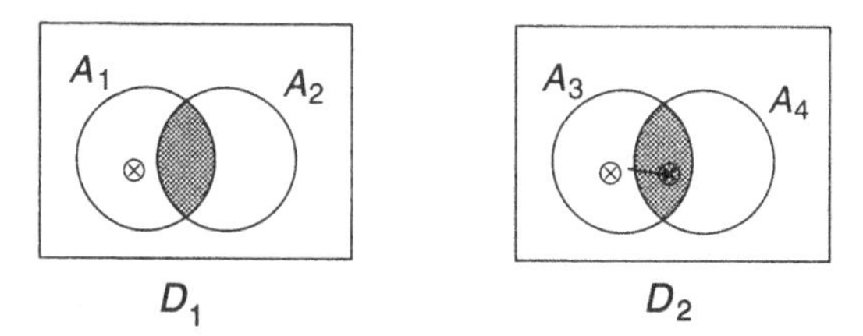

The region of D_2 that is equivalent to $A_1 - A_2$ of D_1 (i.e., $A_3 - A_4$) does not have an x-sequence. Thus, it is not the case that $D_1 \subseteq_{cp} D_2$. The region of D_1 that is equivalent to A_3 of D_2 (i.e., A_1) does not have an x-sequence. Thus, it is not the case that $D_2 \subseteq_{cp} D_1$ either.

Example 16 *Let $\langle A_1, A_2 \rangle \in cp$.*

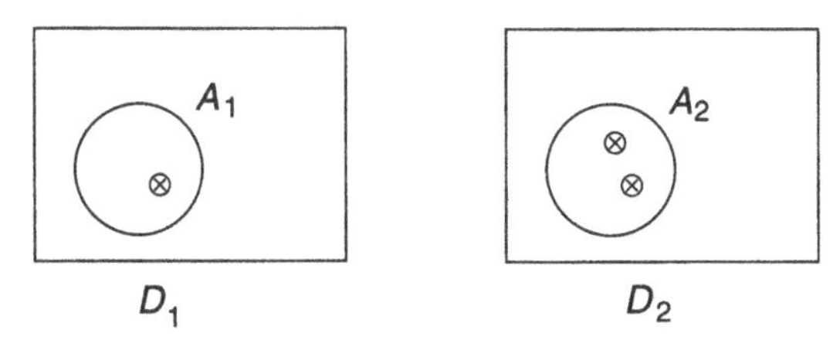

Despite the fact that A_1 has one $\otimes^1$ and A_2 has two $\otimes^1$'s, by definition, $D_1 \equiv_{cp} D_2$.

3.4.2 Equivalent representing facts

In many cases, we need to know whether given diagrams convey the same information or not. A syntactic equivalence between diagrams defined in the previous subsection is too fine-grained for this need. That is, if two diagrams are syntactically equivalent, then by the definition of the set of representing facts of a diagram (i.e., $RF(D)$ defined in §3.3.2), they convey the same information. However, a syntactic equivalence is not a necessary condition for a semantic equivalence.[15]

Before the definition of an equivalence relation between diagrams, we defined an equivalence relation between sets of regions. Similarly, before defining an equivalence relation between the semantic contents of diagrams, we will define an equivalence relation between representing infons.

In the following, we want to say that the two diagrams represent some information in common.

Example 17 *Let* $\langle A_1, A_3 \rangle$, $\langle A_2, A_4 \rangle \in cp$.

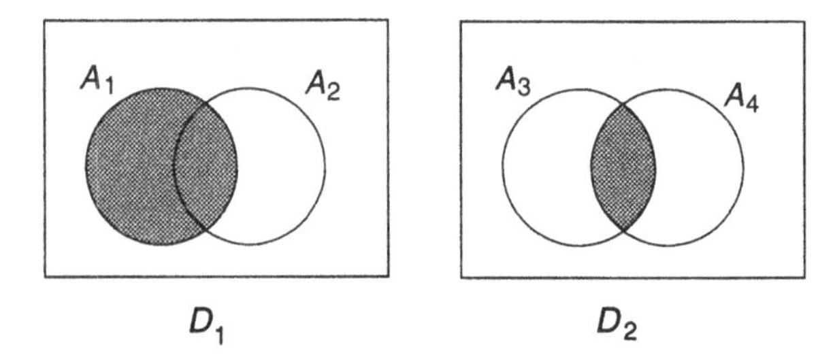

Since two regions, A_1-*and*-A_2 and A_3-*and*-A_4, are *cp*-related to each other, we know that for any set assignment s, s satisfies $\langle\langle \text{Shading}, A_1\text{-}and\text{-}A_2; 1 \rangle\rangle$ if and only if s satisfies $\langle\langle \text{Shading}, A_3\text{-}and\text{-}A_4; 1 \rangle\rangle$. That is, even though $\langle\langle \text{Shading}, A_1\text{-}and\text{-}A_2; 1 \rangle\rangle$ and $\langle\langle \text{Shading}, A_3\text{-}and\text{-}A_4; 1 \rangle\rangle$ are two different representing facts, both of them are always supported by the same set assignments. Let us call these infons equivalent representing facts. Some properties of equivalent representing facts will be used conveniently in later proofs. After the following formal definition of equivalent representing facts, we will prove some of their properties.

Let x_1 and x_2 be representing infons. Then, we define equivalent infons as follows:

$x_1 \equiv_{cp} x_2$ iff two following conditions are satisfied:

(i) either $x_1 = \langle\langle \text{Shading}, A_1; 1 \rangle\rangle$ and $x_2 = \langle\langle \text{Shading}, A_2; 1 \rangle\rangle$
 or $x_1 = \langle\langle \otimes^m, A_1; 1 \rangle\rangle$ and $x_2 = \langle\langle \otimes^n, A_2; 1 \rangle\rangle$;

(ii) $\langle A_1, A_2 \rangle \in \overline{cp}$.

[15] Refer to Example 10.

Example 18 *Let* $\langle A_1, A_2\rangle \in cp$. *According to the previous definition,* $\langle\langle\otimes^1, A_1; 1\rangle\rangle \equiv_{cp} \langle\langle\otimes^4, A_2; 1\rangle\rangle$:

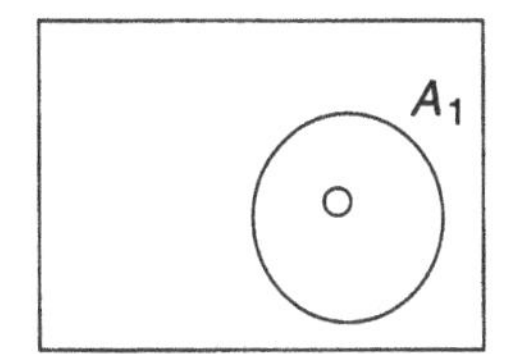

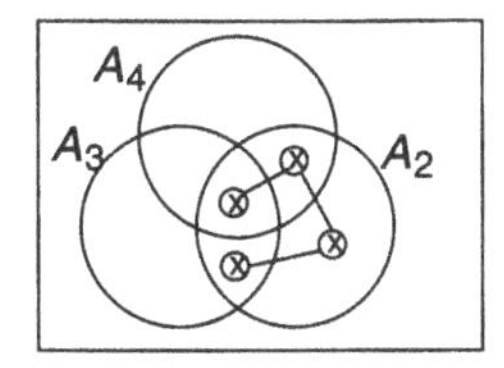

Extending the notion of equivalent representing facts, we can compare the sets of the representing facts of given diagrams. Back in Example 17, every representing fact of D_2 has its equivalent representing fact in D_1, but not vice versa. In Example 18, each diagram has only one representing fact and they are equivalent to each other. We want to define these relations among the representing facts of diagrams as follows:

Let D_1 and D_2 be *wfds*. Let $RF(D_1)$ be the set of representing facts of D_1 and $RF(D_2)$ be the set of representing facts of D_2. We define the following relations between the representing facts of these two diagrams:

$RF(D_1) \subseteq_{cp} RF(D_2)$ iff $\forall_{x \in RF(D_1)} \exists_{y \in RF(D_2)} (x \equiv_{cp} y)$.

$RF(D_1) \equiv_{cp} RF(D_2)$ iff $RF(D_1) \subseteq_{cp} RF(D_2)$ and $RF(D_2) \subseteq_{cp} RF(D_1)$.

$RF(D_1) \subset_{cp} RF(D_2)$ iff $RF(D_1) \subseteq_{cp} RF(D_2)$ and $RF(D_1) \not\equiv_{cp} RF(D_2)$.

Obviously, if $RF(D_1) \subseteq_{cp} RF(D_2)$, then by the definition of "$\models$," $\{D_2\} \models D_1$. Let us go back to some examples in the previous subsection, Examples 12–16: In Example 12, there is no representing fact in D_1 or D_2. Therefore, $RF(D_1) \equiv_{cp} RF(D_2)$. In Example 13, $RF(D_1) \subset_{cp} RF(D_2)$, since there is no representing infon in D_1 equivalent to $\langle\langle$Shading, A_2-*and*-A_3; 1$\rangle\rangle$ which is in $RF(D_2)$. Example 14 is interesting in that $RF(D_1) \equiv_{cp} RF(D_2)$ despite the fact that $D_1 \not\equiv_{cp} D_2$. In Example 15, there is no inclusion relation between the sets of representing facts of these two diagrams. Obviously, $RF(D_1) \equiv_{cp} RF(D_2)$ in Example 16. However, it should be noticed that $\{D_1\} \models D_2$ and $\{D_2\} \models D_1$ in every example from Examples 12 to 16.

In the previous subsection, we defined the equivalence and the inclusion relations among diagrams based upon syntax only. Even if two diagrams have equivalent representing facts, they might not be syntactically equivalent to each other (cf. Example 12 and Example 14). However, if two diagrams are syntactically equivalent, their representing facts should be equivalent to each other. The following propositions are about these two kinds of relations. I will leave out the details of the proofs.

Theorem 6 *If $D' \subseteq_{cp} D$, then $RF(D') \subseteq_{cp} RF(D)$.*

Theorem 7 *If $RF(D_1) \subseteq_{cp} RF(D_2)$ and $BR(D_1) \subseteq_{cp} BR(D_2)$, then $D_1 \subseteq_{cp} D_2$.*

3.5 Transformation

In this section, I aim to define what it is to obtain a diagram from some other diagrams. We want to incorporate the following features for the account of obtainability of a diagram from some given diagrams:

(1) We may always obtain a diagram that is equivalent to one of the given diagrams.
(2) We may always obtain one diagram from others by transformation.

In the case of first-order logic, formula α is derivable from set of formulas Γ if and only if there is a sequence of formulas such that each member of the sequence is *either* a member of Γ or the set of logical axioms *or* the result of applying an inference rule to previous formulas. We will define the obtainability of a diagram from other diagrams in a pretty similar way.

Each deductive system has its own axioms and its own rules of inference to allow us to obtain a desired formula from given formulas. We have a choice between more logical axioms with fewer inference rules and fewer logical axioms with more rules. Let me introduce some terminology: set of tautological diagrams (say, set Ω) and rules of transformation. Set Ω corresponds to the set of logical axioms in a deductive system of first-order logic and the transformation rules are similar to the inference rules. Similarly, one can have a bigger Ω and fewer transformation rules.[16] Here, we take the other route: Choose the smallest set, the empty set, as Ω and have more rules. A process of transformation in this system is analogous to a process of derivation in deductive systems. The first subsection stipulates the rules of transformation, and with these rules

[16]The following is one example for this choice:

Ω is the smallest set of *wfd*s satisfying the following:
1. Any rectangle drawn in the plane is in set Ω.
2. If D is in the set Ω, then if D' results by adding a closed curve interior to the rectangle of D by the partial-overlapping rule, then D' is in set Ω.

This choice does not have R4, the rule of introduction of a basic region, as one of the transformation rules.

the second subsection defines formally what it is for one diagram to be obtainable from other diagrams in this system.

3.5.1 Rules of transformation

We introduce the following six transformation rules into this system:

(1) The rule of erasure of a diagrammatic object.
(2) The rule of erasure of part of an x-sequence.
(3) The rule of spreading x's.
(4) The rule of introduction of a basic region.
(5) The rule of conflicting information.
(6) The rule of unification of diagrams.

Recall that we needed an extra mechanism for tokens of the closed curve and for tokens of the rectangle – a counterpart relation – in the syntactic aspect of the Venn-I system, unlike with symbolic logic systems. This counterpart relation plays an important role in the process of transforming diagrams, which also takes place on the syntactic level.

Suppose that by applying a rule of inference, *modus ponens*, we derived A_2 from two premises: $(A_1 \rightarrow A_2)$ and A_1. *Modus ponens* was applied in the following way: The second premise and the antecedent of the first premise are tokens of the same sentence-symbol type. Hence, we can *copy* the consequence of the first premise, that is, A_2. This is the same sentence-symbol type as the consequence of the first premise. If we copy a token of a symbol type, then the new token belongs to the same symbol type as the original token.

In Venn-I, if we copy a token of a closed curve, the new closed-curve token is not only a token of the same type (i.e., the closed-curve type) but also a token that is equivalent to the original token of a closed curve. The same thing goes for rectangle tokens. That is, if a user *copies* a token of a diagrammatic-object type, then he chooses a counterpart relation that includes the ordered pair of the original token and the new token.

Definition *Let A and B be both rectangles or both closed curves. Then, if B is a copy of A, then* $\langle$*region A, region B*$\rangle \in cp$.

Definition *Let* D_1 *and* D_2 *be wfds. If* D_2 *is a copy of* D_1*, then*

(i) *every rectangle and every closed curve of* D_1 *is copied in* D_2*,*
(ii) *every shading in some region of* D_1 *is drawn in the* $\overline{cp}$*-related region of* D_2*,*

(iii) *every x-sequence in some region of* D_1 *is drawn in the* $\overline{cp}$*-related region of* D_2,
(iv) D_2 *has nothing else.*

Therefore:

Theorem 8 *Let* D_1 *and* D_2 *be wfds. If* D_2 *is a copy of* D_1, *then* $D_1 \equiv_{cp} D_2$.

I will assume this property of "being a copy of" for the specification of each transformation rule.

Before we move on to the explanation of each rule of transformation, there is one point worth mentioning. We are concerned with transformations only from well-formed diagrams to well-formed diagrams. So we can assume that given diagrams are always well-formed. We also have to make sure that the derived diagram from those given diagrams will be a well-formed diagram. Accordingly, each rule should be formulated in such a way as to prevent us from getting any ill-formed diagrams.

Rule 1: The rule of erasure of a diagrammatic object

We may copy a *wfd* omitting one of the following diagrammatic objects: an entire shading in a minimal region, a whole x-sequence, or a closed curve. When a closed curve is erased, (i) if a shading remains partially in a minimal region (i.e., a shading does not fill up a minimal region), then the shading should be erased, and (ii) if an x-sequence remains with more than one ⊗ in the same minimal region, then that part of the x-sequence (i.e., more than one ⊗ in the same minimal region) should be replaced with one ⊗ and should be connected with the rest of the x-sequence.

Let us go through examples for erasing each object.

(1) When we omit the shading in some region, we should erase the entire shading in a minimal region. Otherwise, we would get an ill-formed diagram.

Example 19 *We are allowed to transform the diagrams on the left to the diagrams on the right, by applying this rule three times:*

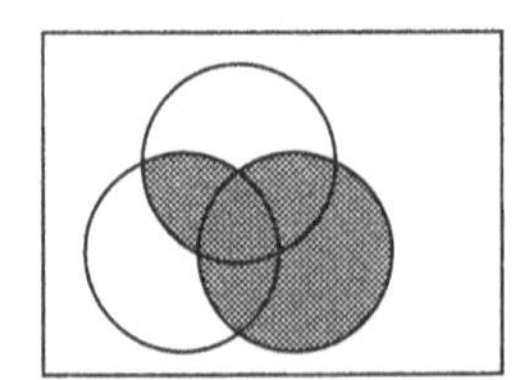 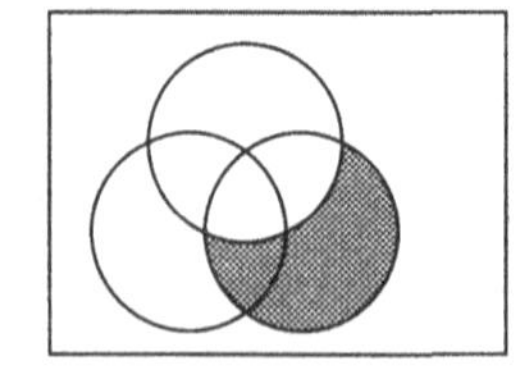

(2) The erasure of a *whole* x-sequence allows the transformation from the left figure to the right one.

Example 20 *We erase* $\otimes^3$ *from* D *to get* D_1 *and erase* $\otimes^1$ *from* D_1 *to get* D_2*:*

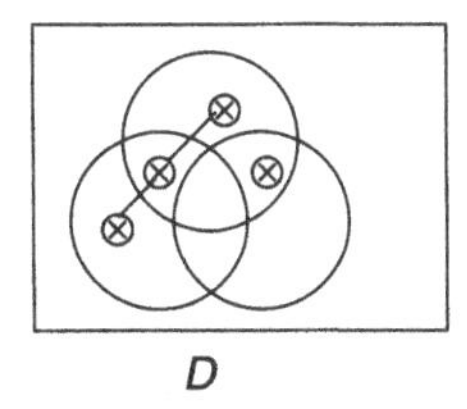

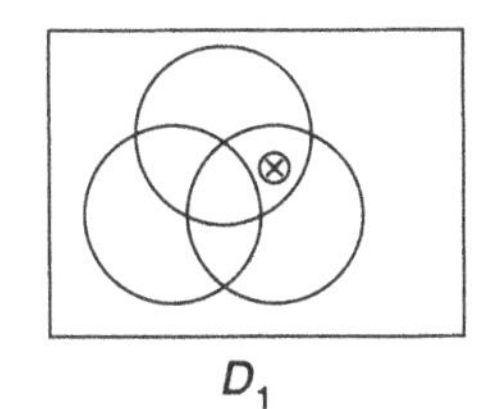

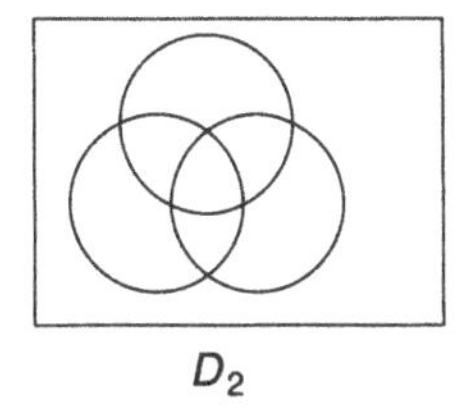

(3) When we erase a closed curve, certain regions disappear. A shading or an x-sequence on these regions might have to be erased or be modified as the rule specifies. Otherwise, the resulting diagram might be ill-formed.

Example 21 *In the following,* D_1 *is the result of erasing closed curve* B *from* D *and* D_2 *is what we get after erasing* A *from* D*:*

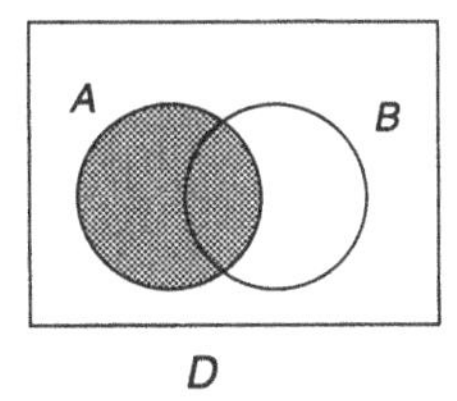

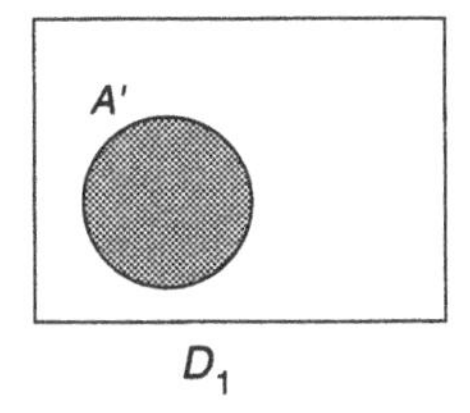

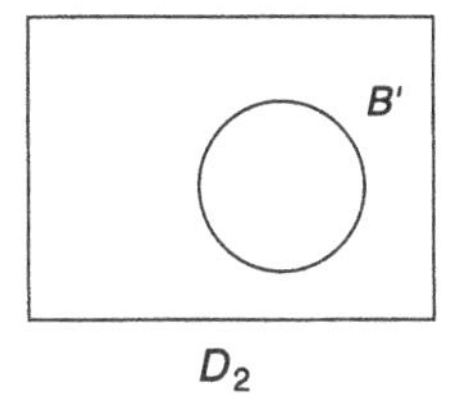

When closed curve B *is erased, the shading that remains fills up minimal region* A'*, but when* A *is erased the following results:*

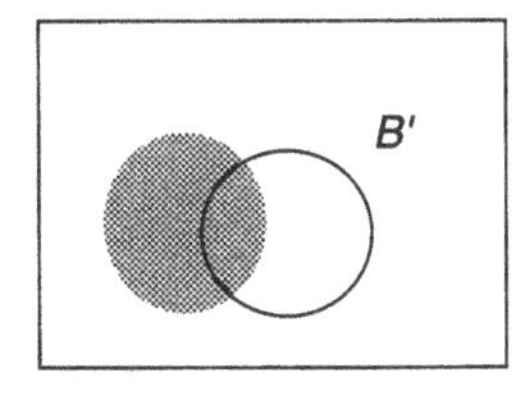

These shadings do not fill up any minimal region. Rule 1 requires that these shadings should be erased to get diagram D_2*.*

Example 22 *In this example, whether we erase* A *or* B *in* D*, region* A*-and-*B *is erased. After one of the closed curves is erased, the shading will not fill up the region any more. That is, the resulting diagram would be* *ill-formed. Accordingly, the shading on A-and-B should be erased, so that the resulting diagram remains well-formed.*

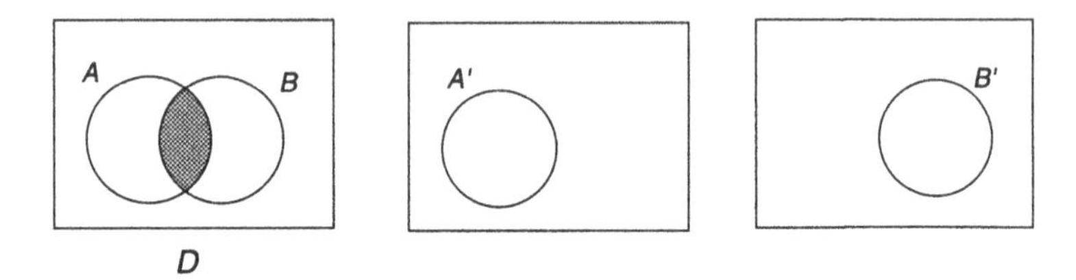

Example 23 *In the following, we want to erase closed curve B from the diagram on the left to get the diagram on the right.*

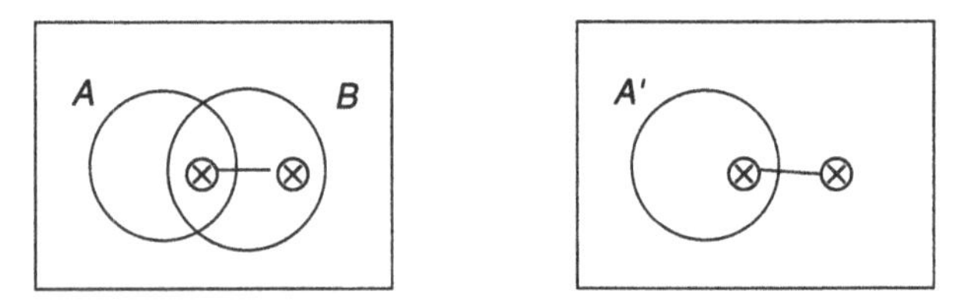

In this new diagram, the x-sequence has each ⊗ in different minimal regions. Accordingly, this is a resulting diagram as it is.

Example 24 *Suppose we erase the closed curve from D to get the middle diagram. However, in the middle one, we find two ⊗'s of the x-sequence in the same minimal region. Rule 1 says that we should replace these two with one ⊗ to get the third one:*

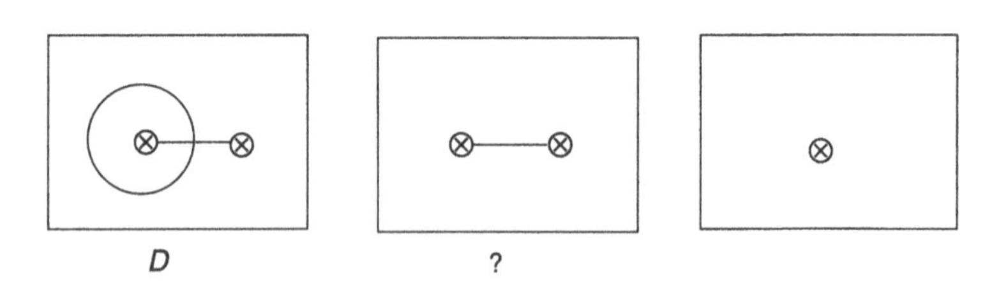

Example 25 *In the following, D_1 is the result of erasing closed curve B from D. Since two ⊗'s are in minimal region A', we replace that part with one ⊗ and connect it with the rest of the x-sequence to get the third diagram, D_2:*

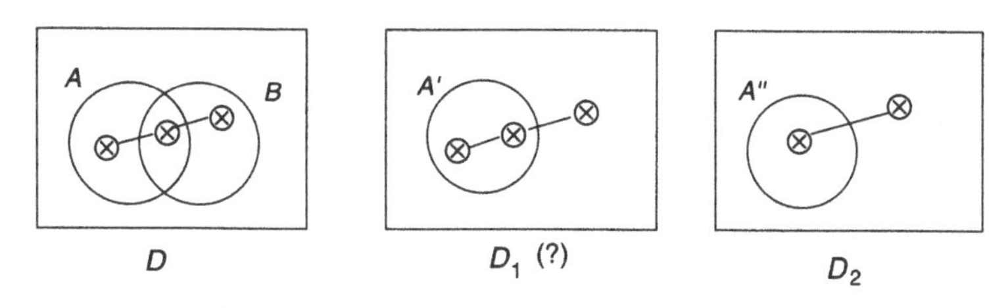

Rule 2: The rule of erasure of part of an x-sequence

We may copy a *wfd* omitting any subpart of an x-sequence if that part is in a shaded region. The number of x-sequences in a diagram does not increase. That is, if ⊗ in a shaded region is at the end of an x-sequence, we may erase -⊗ or ⊗- so that the remaining part is one x-sequence. If ⊗ in a shaded region is in the middle of an x-sequence, after erasing this ⊗ we should connect both of the remaining parts so that we have one x-sequence (not two x-sequences by disconnection).

Example 26 *Let us compare the following diagrams:*

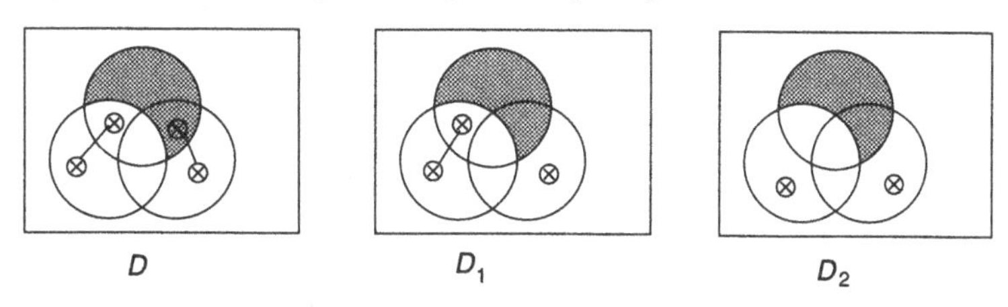

The transformation from D to D_1 is legitimate by this rule. However, we do not have any rule to allow the transformation from D_1 to D_2. The part of the x-sequence that is erased in the diagram D_2, that is, ⊗-, is not in the shaded part of D_1. Therefore, rule 2 does not allow this partial erasure. Rule 1 is concerned only with the erasure of a whole x-sequence, not with a proper subpart of an x-sequence.

Example 27 *Let us compare the following diagrams:*

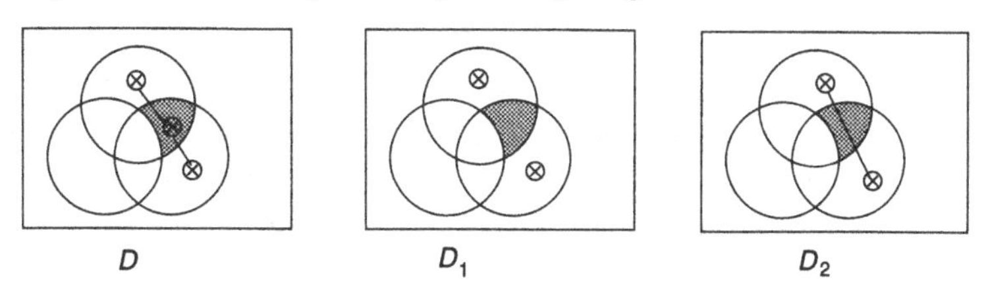

The transformation from D to D_1 is not allowed, since D has one x-sequence whereas D_1 has two. After ⊗ in the shaded region of D is erased, the remaining parts should be connected as in diagram D_2, as rule 2 specifies.

Rule 3: The rule of spreading x's

If *wfd D* has an x-sequence, then we may copy *D* with ⊗ drawn in some *other* minimal region and connected to the existing x-sequence.

Example 28 *In the following example, by applying this rule twice we get the rightmost diagram from the leftmost one:*

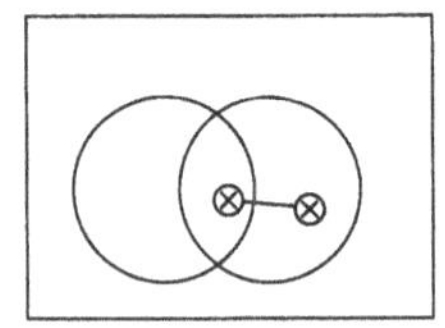
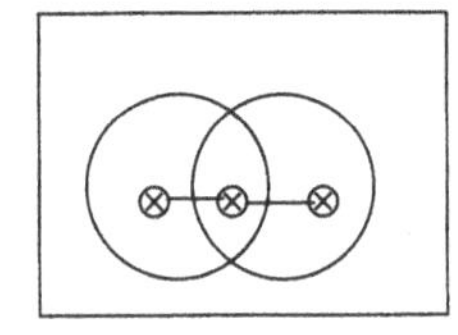
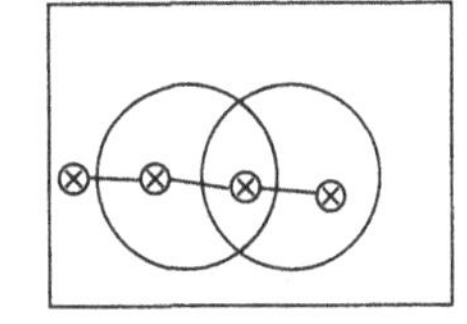

This rule has an important condition about "other minimal regions" for an extended part of an x-sequence. Otherwise, we would get ill-formed diagrams as follows:

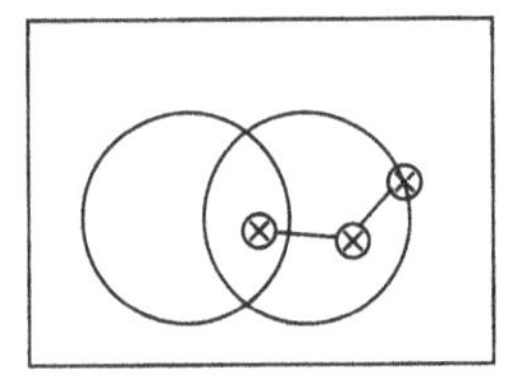
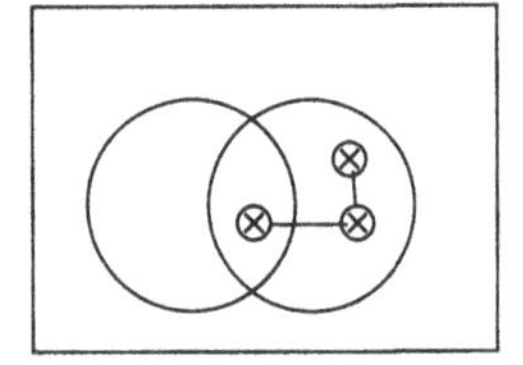

Rule 4: The rule of introduction of a basic region

A basic region may be introduced by drawing either a rectangle or a closed curve. Since a *wfd* has only one rectangle, we may introduce a rectangle only if no *wfd* is given. Otherwise, we may copy a *wfd* to introduce a basic region by adding a new closed curve to a given diagram in the following way: (i) A new closed curve should be drawn interior to the rectangle observing the partial-overlapping rule (described in rule 2 of the well-formedness rules). (ii) If there is an x-sequence in the original diagram, then each ⊗ of an x-sequence is replaced with ⊗–⊗ (where each ⊗ is placed in a new minimal region). The number of x-sequences does not change. (I.e., suppose that an x-sequence, say, $\otimes^k$, is in $A_1 + \cdots + A_k$

(where $A_1, \ldots, A_k$ are minimal regions). With the introduction of a new closed curve, each $A_i (1 \leq i \leq k)$ is divided into two minimal regions, say, A_i' and A_i''. Therefore, in a new diagram, x-sequence $\otimes^{2k}$ is placed in $A_1' + A_1'' + \cdots + A_k' + A_k''$ (where $A_1', \ldots, A_k''$ are minimal regions).)

Example 29 *We add a new closed curve interior to the rectangle, observing the partial-overlapping rule.*

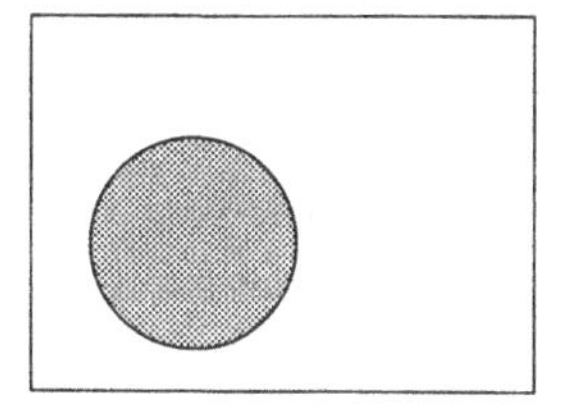

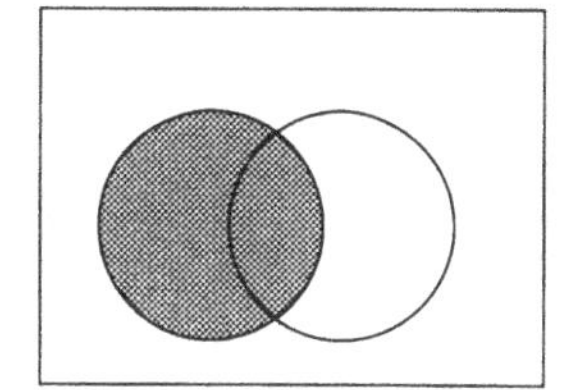

Example 30 *In the following, adding closed curve C causes some change in minimal regions. Region A − B is divided into two minimal regions, (A − B) − C and (A − B)-and-C, and similarly for region A-and-B (i.e., into (A-and-B) − C and (A-and-B)-and-C). Rule 4 requires that each ⊗ of these two minimal regions in the original diagram should be replaced with ⊗–⊗. Importantly, as the original x-sequence is connected, these revised parts should be connected with each other so that the number of x-sequences does not change at all.*

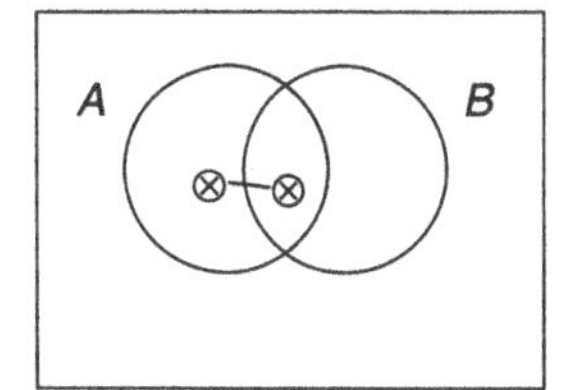

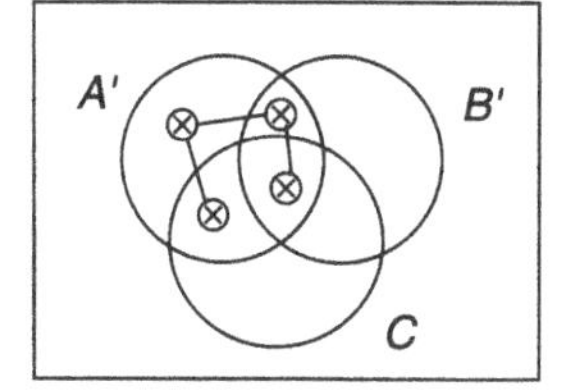

Rule 5: The rule of conflicting information

If a diagram has a region with both a shading and an x-sequence, then we may transform this diagram to any diagram.

Example 31 *Region $A - B$ of the diagram on the left has both a shading and an x-sequence, $\otimes^1$. Therefore, this rule allows us to get the diagram on the right:*

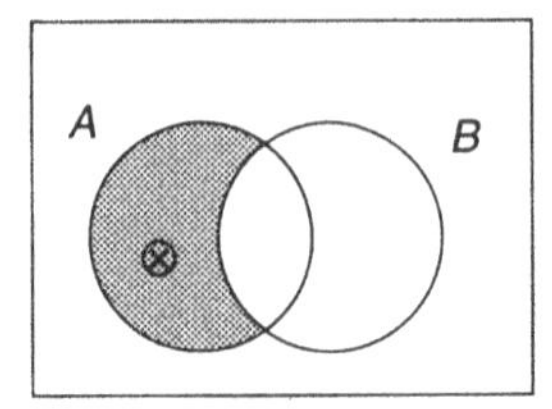

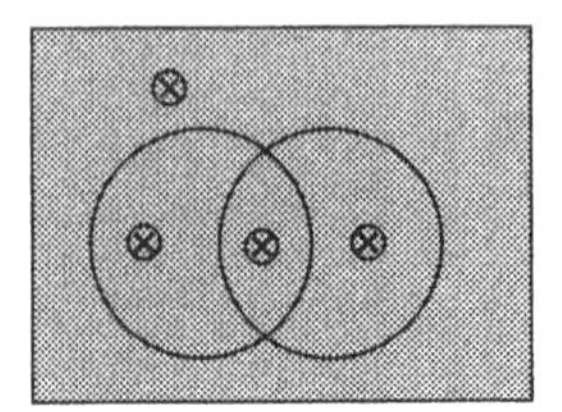

Example 32 *Region A of the diagram on the left has both a shading and an x-sequence $\otimes^2$. Therefore, this rule allows us to get the diagram on the right:*

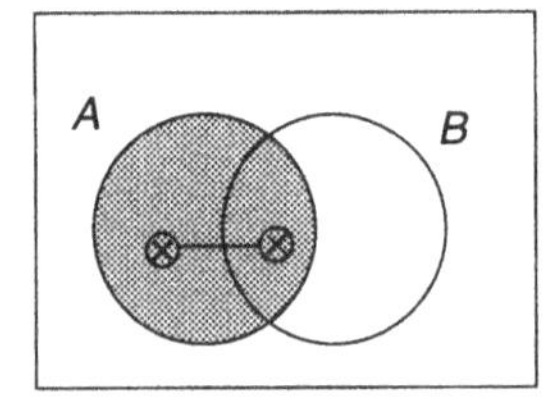

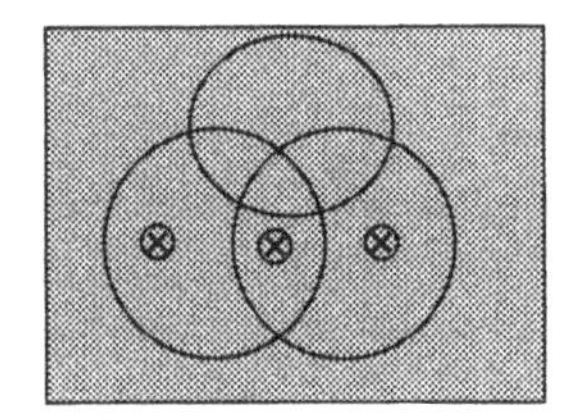

Rule 6: The rule of unification of diagrams

We may unify two diagrams, D_1 and D_2, into one diagram, call it D, if a given *cp* relation contains the ordered pair of the rectangle of D_1 and the rectangle of D_2.

D is the *unification* of D_1 and D_2 if the following conditions are satisfied:

(1) The rectangle and the closed curves of D_1 are copied in D.
(2) The closed curves of D_2 that do not stand in the given *cp* relation to any of the closed curves of D_1 are copied in D interior to the rectangle observing the partial-overlapping rule.
(3) For any region A shaded in D_1 or D_2, the $\overline{cp}$-related region of D should be shaded.
(4) For any region A with an x-sequence in D_1 or D_2, an x-sequence should be drawn in the $\overline{cp}$-related regions of D, satisfying the well-formedness rules of diagrams.

Let me illustrate this rule through several examples.

Example 33 *Two diagrams D_1 and D_2 are given, where $\langle U_1, U_2\rangle$, $\langle A_1, A_2\rangle$ $\in$ cp:*

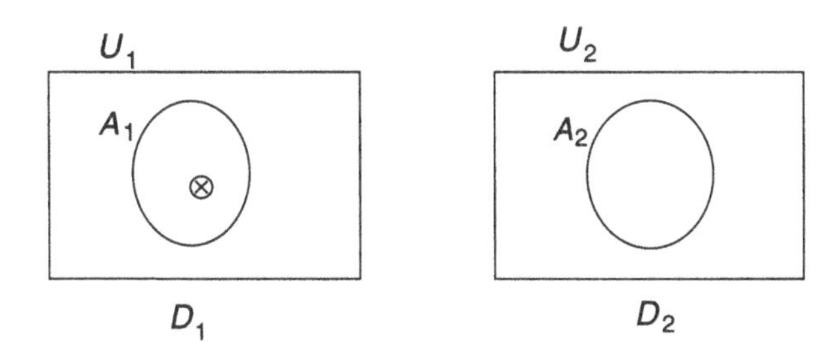

Since the given cp relation holds between the rectangle of D_1 and the rectangle of D_2, we can unify the two diagrams. First, we copy the rectangle and the closed curve of D_1 and name them U_3 and A_3, respectively. Accordingly, $\langle U_1, U_3\rangle$, $\langle A_1, A_3\rangle \in cp$. Since the closed curve of D_2, that is, A_2, is cp-related to the closed curve of D_1, that is, A_1, we do not add any closed curve. The x-sequence in region A_1 of D_1 should be drawn in region A_3 of D, since A_1 and A_3 are $\overline{cp}$-related and A_3 is a minimal region. Therefore, this rule allows us to unify diagrams D_1 and D_2, giving diagram D:

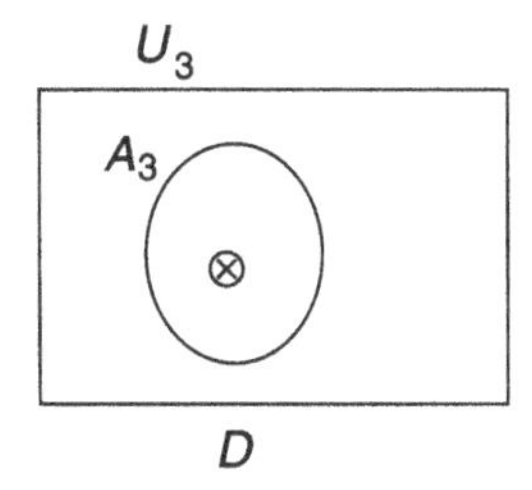

Example 34 *Let $\langle U_1, U_2\rangle \in cp$, and $\langle A_1, A_2\rangle \notin cp$. Since A_2 is not cp-related to any closed curve in D_1, we draw the cp-related closed curve, that is, A_4, in D, to get diagram D as follows (notice that the partial-overlapping rule is observed):*

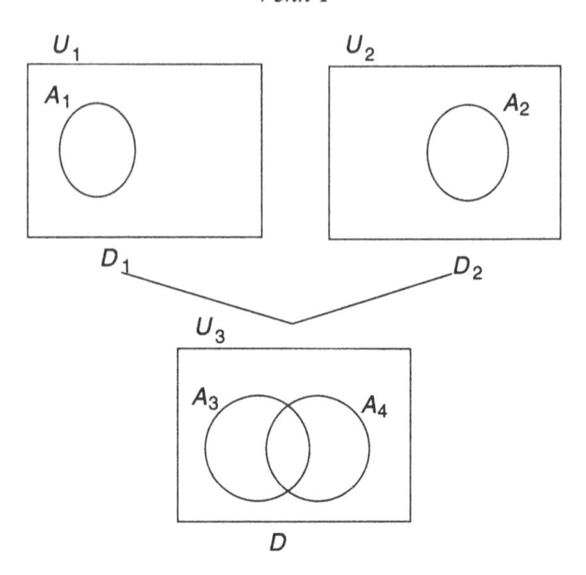

Example 35 *Let* $\langle U_1, U_2 \rangle \in cp$, $\langle A_2, A_4 \rangle \in cp$, $\langle A_1, A_3 \rangle \notin cp$. *After copying the rectangle and the closed curves of* D_1, *we need to copy closed curve* A_3. *Therefore, the following holds:* $\langle U_1, U_3 \rangle$, $\langle U_2, U_3 \rangle$, $\langle A_1, A_6 \rangle$, $\langle A_2, A_7 \rangle$, $\langle A_3, A_5 \rangle$, $\langle A_4, A_7 \rangle \in cp$.

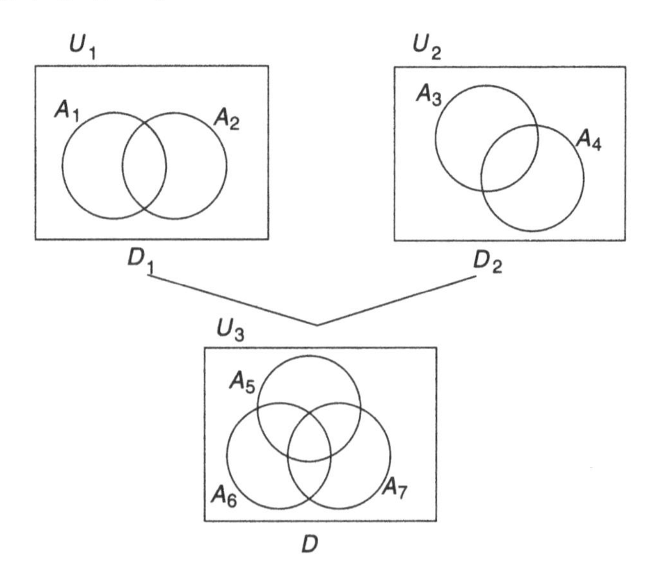

Now, we will see how this rule transforms diagrams with shadings or x-sequences into one diagram:

Example 36 *Let $\langle U_1, U_2 \rangle \in cp$ and $\langle A_1, A_3 \rangle$, $\langle A_2, A_3 \rangle \notin cp$.*

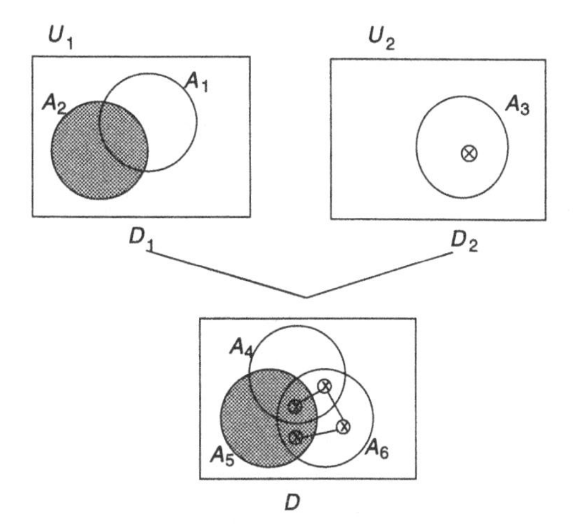

Suppose that closed curve A_4 is the copy of A_1, A_5 is that of A_2 and A_6 is that of A_3. Accordingly, $\langle A_1, A_4 \rangle$, $\langle A_2, A_5 \rangle$, $\langle A_3, A_6 \rangle \in cp$. Clause 3 of the unification rule tells us that the shading in region A_2 of D_1 should be copied in the $\overline{cp}$-related region, that is, region A_5, in diagram D. Clause 4 of the unification rule and our well-formedness rules guide us to draw $\otimes^4$ in the $\overline{cp}$-related region, that is, A_6, to get a well-formed unified diagram D. Let's look at two more examples:

Example 37 *Let $\langle U_1, U_2 \rangle$, $\langle A_2, A_4 \rangle \in cp$ and $\langle A_1, A_3 \rangle \notin cp$.*

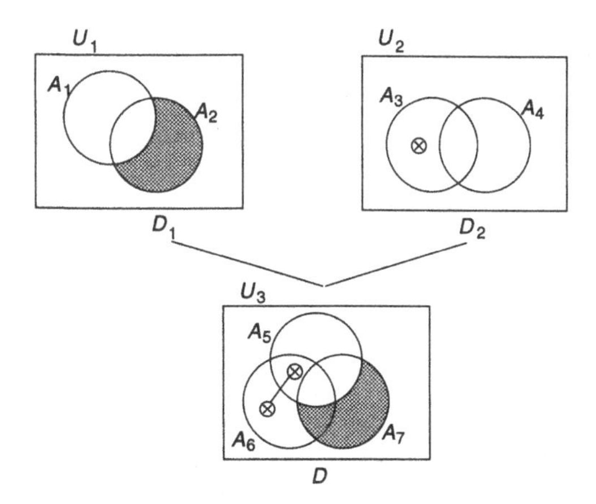

Example 38 *Let* $\langle U_1, U_2\rangle$, $\langle A_1, A_4\rangle$, $\langle A_2, A_5\rangle$, $\langle A_3, A_6\rangle \in cp$.

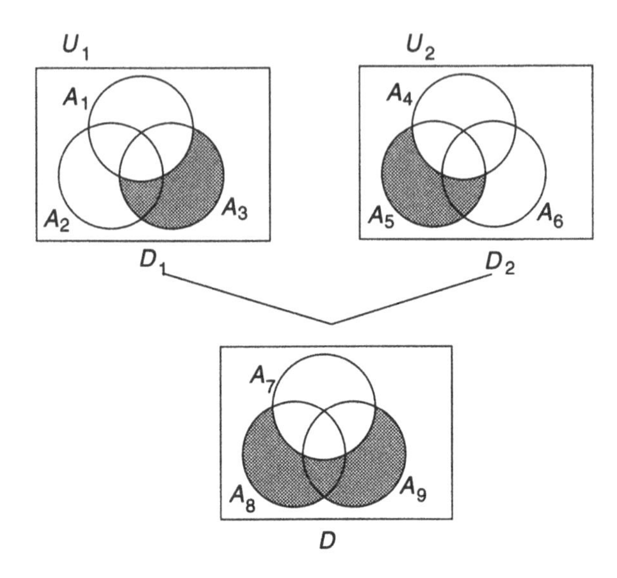

3.5.2 *Formal obtainability*

With these six transformation rules, we define obtainability of a diagram from given diagrams.

Definition *Let* Δ *be an empty set or a set of one wfd or a set of two wfds. Then,* $\Delta \rightsquigarrow D$ *iff there is a rule of transformation such that it allows us to transform the diagrams of* Δ *to* D.

As we discussed §3.4.1, in the case of Venn-I the comparison of diagrams is more complicated than in the case of a sentential or a first-order language. We needed to define the equivalence among diagrams. With this definition, we can define the obtainability of a diagram from given diagrams in the following way:

Definition *Let* Δ *be a set of wfds and* D *be a wfd. Wfd* D *is obtainable from a set* Δ *of wfds* ($\Delta \vdash D$) *iff there is a sequence of wfds* $\langle D_1, \ldots, D_n\rangle$ *such that* $D_n \equiv_{cp} D$ *and for each* k *(where* $1 \le k \le n$*)* <u>*either*</u>

(a) *there is some* D' *such that* $D' \in \Delta$ *and* $D' \equiv_{cp} D_k$, <u>*or*</u>
(b) *there is a set of diagrams* Δ' *such that for every diagram* D_i *of* Δ', $i < k$, *and* $\Delta' \rightsquigarrow D_k$.

We can prove the following propositions, which will be used conveniently in our further discussion. One says that for a given diagram, we can always obtain its subdiagrams by transformation. The other says that for a given diagram we can always obtain any diagram the set of whose representing facts is a subset of the set of the representing facts of the given diagram.

Theorem 9 *If* $D' \subseteq_{cp} D$, *then* $\{D\} \models D'$ *and* $\{D\} \vdash D'$.

Proof This follows by Theorem 6 and by the rule of erasure of a diagrammatic object. □

Theorem 10 *If* $RF(D') \subseteq_{cp} RF(D)$, *then* $\{D\} \vdash D'$.

Proof Refer to the appendix.

3.6 Soundness

We defined what it is for one diagram to follow from other diagrams ($\Delta \models D$). We also defined what it is for one diagram to be obtained from other diagrams ($\Delta \vdash D$). Now, we raise the question of the soundness of this representation system. Whenever one *wfd* D is obtainable from a set Δ of *wfd*s (i.e., $\Delta \vdash D$), is it the case that D follows from Δ (i.e., $\Delta \models D$)? The main idea is to prove that all the rules of transformation are valid. The first subsection will be devoted to this proof and we will complete the soundness proof in the second subsection.

3.6.1 Validity of rules of transformation

In this section I prove that each transformation rule is valid. That is,

$$\text{if } \Delta \leadsto D, \text{ then } \Delta \models D.$$

Rule 1: The rule of erasure of a diagrammatic object

Let $\{D_1\} \leadsto D$, by the rule of erasure of a diagrammatic object.

(i) It is valid when a closed curve is erased.

Since we erased a closed curve, $BR(D) \subset_{cp} BR(D_1)$. That is, every region in D has its equivalent region in D_1. Suppose that $\{D_1\} \not\models D$. This means that there is $\alpha \in RF(D)$ such that $\exists_s(s \models D_1 \wedge s \not\models \alpha)$. Let s_1 be such that $s_1 \models D_1$ and $s_1 \not\models \alpha$.

(1) Let $\alpha = \langle\langle \text{Shading}, A, 1\rangle\rangle$. Accordingly, $\overline{s_1}(A) \neq \emptyset$. Let A_1 be a region of D_1 such that $\langle A_1, A\rangle \in \overline{cp}$. Hence, $\overline{s_1}(A_1) \neq \emptyset$. But, since we did not add a shading to D, it must be the case that $\langle\langle \text{Shading}, A_1; 1\rangle\rangle \in RF(D_1)$. Since $s_1 \models D_1$, $\overline{s_1}(A_1) = \emptyset$. This is a contradiction. (2) Let $\alpha = \langle\langle \otimes^n, A; 1\rangle\rangle$. Hence, $\overline{s_1}(A) = \emptyset$. Accordingly, $\overline{s_1}(A_1) = \emptyset$, where $\langle A_1, A\rangle \in \overline{cp}$. Since we did not add an x-sequence to D, it must be the case that there is a region A_2 in D_1 such that A_2 is a part of A_1 and $\langle\langle \otimes^m, A_2; 1\rangle\rangle \in RF(D_1)$. By Theorem 3, $\overline{s_1}(A_2) \subseteq \overline{s_1}(A_1)$. Since $\overline{s_1}(A_1) = \emptyset$, $\overline{s_1}(A_2) = \emptyset$. Then, $s_1 \not\models \langle\langle \otimes^m, A_2; 1\rangle\rangle$, which contradicts the assumption that $s_1 \models D_1$.

(ii) It is valid when a shading is erased.
Since $RF(D) \subseteq_{cp} RF(D_1)$, $\{D_1\} \models D$.

(iii) It is valid when an x-sequence is erased.
Since $RF(D) \subseteq_{cp} RF(D_1)$, $\{D_1\} \models D$.

Rule 2: The rule of the erasure of part of an x-sequence

Suppose that $\{D_1\} \rightsquigarrow D$, where in D_1 an x-sequence, say, $\otimes^n$, is in a region, say, A, where $A = A_1 + \cdots + A_n$ and each A_i(where $1 \leq i \leq n$) is a minimal region with $\otimes$ of the x-sequence. Also suppose that A_k(for some k such that $1 \leq k \leq n$) is shaded. Therefore,

$$\langle\langle \otimes^n, A; 1\rangle\rangle \in RF(D_1), \tag{1}$$

where $A = A_1 + \cdots + A_n$ and A_i $(1 \leq i \leq n)$ is a minimal region with $\otimes$.

$$\langle\langle \text{Shading}, A_k; 1\rangle\rangle \in RF(D_1), \text{ for some } k \text{ such that } 1 \leq k \leq n. \tag{2}$$

In resulting diagram D, the proper subpart of this x-sequence in a shaded region is erased. Therefore,

$$\langle\langle \otimes^{n-1}, B^*; 1\rangle\rangle \in RF(D), \tag{3}$$

where $B^* = B_1 + \cdots + B_{k-1} + B_{k+1} + \cdots + B_n$ and $\langle A_1, B_1\rangle, \ldots, \langle A_n, B_n\rangle \in \overline{cp}$. Let set assignment s_1 be such that

$$s_1 \models \langle\langle \otimes^n, A; 1\rangle\rangle, \tag{4}$$

$$s_1 \models \langle\langle \text{Shading}, A_k; 1\rangle\rangle. \tag{5}$$

By equations (4) and (5),

$$\overline{s_1}(A - A_k) \neq \emptyset. \tag{6}$$

By equation (3),

$$\langle A - A_k, B^*\rangle \in \overline{cp}. \tag{7}$$

By equations (6) and (7),

$$\overline{s_1}(B^*) \neq \emptyset. \tag{8}$$

By equations (4), (5), and (8),

$$\forall_s((s \models \langle\langle \otimes^n, A; 1\rangle\rangle \wedge s \models \langle\langle \text{Shading}, A_k; 1\rangle\rangle) \longrightarrow s \models \langle\langle \otimes^{n-1}, B^*; 1\rangle\rangle). \tag{9}$$

Therefore, $\{D_1\} \models D$.

Rule 3: The rule of spreading x's

Suppose that $\langle\langle \otimes^n, A_1; 1\rangle\rangle \in RF(D_1)$. Draw a new $\otimes$ in some minimal region of D, say, A^*, and connect it to an existing x-sequence of A, where $\langle A_1, A\rangle \in \overline{cp}$. That is, $\langle\langle \otimes^{n+1}, A + A^*; 1\rangle\rangle \in RF(D)$. For any set assignment s, $\overline{s}(A) \subseteq \overline{s}(A + A^*)$. Since $\langle A_1, A\rangle \in \overline{cp}$, $\overline{s}(A_1) \subseteq \overline{s}(A + A^*)$. Accordingly, $\forall_s(s \models \langle\langle \otimes^n, A_1; 1\rangle\rangle \longrightarrow s \models \langle\langle \otimes^{n+1}, A + A^*; 1\rangle\rangle)$. Therefore, $\{D_1\} \models D$.

Rule 4: The rule of introduction of a basic region

There are two cases. One is when no *wfd* is given. In this case, this rule allows us to draw a rectangle. Therefore, for this new diagram D, $RF(D) = \emptyset$. Trivially, $\Delta \models D$ when $\Delta = \emptyset$. The other case is that $\{D_1\} \rightsquigarrow D$, where we introduce a basic region by adding a closed curve, say, B, to a diagram D_1.

If there is no representing fact in D_1, then D does not have any representing fact either, since adding a closed curve does not add a new representing fact. Therefore, trivially $\{D_1\} \models D$.

Suppose that $\langle\langle \text{Shading}, A_1, 1\rangle\rangle \in RF(D_1)$. By the partial-overlapping rule, a new closed curve overlaps A partially, where $\langle A_1, A\rangle \in \overline{cp}$. Accordingly, there are new regions in D which do not have any equivalent region in D_1: *A-and-B*, $A + B$, $A - B$, and $B - A$. Among them, *A-and-B* and $A - B$ are the shaded region. Therefore, $\langle\langle \text{Shading}, A\text{-and-}B; 1\rangle\rangle$ and $\langle\langle \text{Shading}, A - B; 1\rangle\rangle$ are in $RF(D)$ but do not have an equivalent representing fact in D_1. But, $\forall_s(s \models \langle\langle \text{Shading}, A_1, 1\rangle\rangle \longrightarrow (s \models \langle\langle \text{Shading}, A\text{-and-}B; 1\rangle\rangle \wedge s \models \langle\langle \text{Shading}, A - B; 1\rangle\rangle))$.

Suppose that $\langle\langle \otimes^n, A_1; 1\rangle\rangle \in RF(D_1)$. Introducing a new closed curve changes the x-sequence because of the change of minimal regions. However, $\langle\langle \otimes^m, A; 1\rangle\rangle \in RF(D)$, where $\langle A_1, A\rangle \in \overline{cp}$. Notice that $\langle\langle \otimes^n, A_1; 1\rangle\rangle \equiv_{cp} \langle\langle \otimes^m, A; 1\rangle\rangle$.

Therefore, $\{D_1\} \models D$.

Rule 5: The rule of conflicting information

Suppose that $\langle\langle \otimes^n, A_1; 1\rangle\rangle \in RF(D_1)$ and $\langle\langle \text{Shading}, A_1; 1\rangle\rangle \in RF(D_1)$. But, there is no set assignment s such that $\bar{s}(A_1) = \emptyset$ and $\bar{s}(A_1) \neq \emptyset$. Therefore, by definition of the consequence relation "$\models$," trivially $\{D_1\} \models D$ for any D.

Rule 6: The rule of unification of diagrams

Let $\{D_1, D_2\} \rightsquigarrow D$, by unification.

(Case 1) Let $\langle\langle \text{Shading}, A; 1\rangle\rangle \in RF(D)$, but there is no equivalent infon either in $RF(D_1)$ or in $RF(D_2)$.

(i) Suppose that there is a region in D_1,[17] say, A_1, that is equivalent to A. Then, it should be the case that $\langle\langle \text{Shading}, A_1; 1\rangle\rangle \notin RF(D_1)$. Otherwise, it would be the case that $\langle\langle \text{Shading}, A_1; 1\rangle\rangle \equiv_{cp} \langle\langle \text{Shading}, A; 1\rangle\rangle$.[18] Therefore, the shading in region A of the unified diagram D must be copied from the shading of some region of D_2, call it B_2, in the following way: By unification, the shading of B_2 is copied into the equivalent region of D, call it B. That is, $\langle B_2, B\rangle \in \overline{cp}$. But, since A is a part of B, A is also shaded in this process. That is,

$$\langle\langle \text{Shading}, B_2; 1\rangle\rangle \in RF(D_2). \tag{1}$$

Since A is a part of B,

$$\forall_s(\bar{s}(A) \subseteq \bar{s}(B)). \tag{2}$$

Since $\langle B_2, B\rangle \in \overline{cp}$,

$$\forall_s(\bar{s}(A) \subseteq \bar{s}(B_2)). \tag{3}$$

By equations (1) and (3),

$$\forall_s(s \models D_2 \longrightarrow s \models \langle\langle \text{Shading, A; 1}\rangle\rangle). \tag{4}$$

(ii) Suppose that there is no equivalent region to A in either D_1 or D_2. Then, A must be one of the following cases:

(a) $A = B_1 + B_2$,
(b) $A = B_1\text{-}and\text{-}B_2$,
(c) $A = B_1 - B_2$,
(d) $A = B_2 - B_1$,

[17] It cannot be the case that both D_1 and D_2 have regions equivalent to A. If it were, there should be an equivalent infon to $\langle\langle \text{Shading}, A; 1\rangle\rangle$. This contradicts our assumption.

[18] It contradicts our assumption that there is no equivalent representing fact to $\langle\langle \text{Shading}, A; 1\rangle\rangle$.

where $\langle A_1, B_1\rangle \in \overline{cp}$, $\langle A_2, B_2\rangle \in \overline{cp}$, A_1 is in D_1, and A_2 is in D_2.[19])

(a) Since $\langle\langle$Shading, $A;1\rangle\rangle \in RF(D)$, $\langle\langle$Shading, $B_1;1\rangle\rangle \in RF(D)$ and $\langle\langle$Shading, $B_2;1\rangle\rangle \in RF(D)$. We know that $\langle\langle$Shading, $A_1;1\rangle\rangle \in RF(D_1)$ and $\langle\langle$Shading, $A_2;1\rangle\rangle \in RF(D_2)$. (Recall the unification process of this rule.) Therefore, we know that $\forall_s((s \models D_1 \ \wedge \ s \models D_2) \longrightarrow (s \models$ $\langle\langle$Shading, $B_1;1\rangle\rangle \wedge s \models \langle\langle$Shading, $B_1;1\rangle\rangle))$. Since $\overline{s}(A) = \overline{s}(B_1) \cup \overline{s}(B_2)$, $\forall_s((s \models D_1 \ \wedge \ s \models D_2) \longrightarrow s \models \langle\langle$Shading, $A;1\rangle\rangle)$.

(b) Since $\langle\langle$Shading, $A;1\rangle\rangle \in RF(D)$, either $\langle\langle$Shading, $B_1;1\rangle\rangle \in RF(D)$ or $\langle\langle$Shading, $B_2;1\rangle\rangle \in RF(D)$. Suppose that $\langle\langle$Shading, $B_1;1\rangle\rangle \in RF(D)$. If $\langle\langle$Shading, $A_1;1\rangle\rangle \in RF(D_1)$, then $\forall_s(s \models D_1 \longrightarrow s \models \langle\langle$Shading, $B_1;1\rangle\rangle)$. Otherwise, by (i), $\forall_s(s \models D_2 \longrightarrow s \models \langle\langle$Shading, $B_1;1\rangle\rangle)$. In either case, $\forall_s((s \models D_1 \ \wedge \ s \models D_2) \longrightarrow s \models \langle\langle$Shading, $B_1;1\rangle\rangle)$. Since $\overline{s}(A) \subseteq \overline{s}(B_1)$, $\forall_s((s \models D_1 \ \wedge \ s \models D_2) \longrightarrow s \models \langle\langle$Shading, $A;1\rangle\rangle)$.

(c) and (d) are similar to case (b).

(Case 2) Let $\langle\langle \otimes^n, A;1\rangle\rangle \in RF(D)$, but there is no equivalent infon to $\langle\langle \otimes^n, A;1\rangle\rangle$ either in $RF(D_1)$ or $RF(D_2)$. There must be representing fact $\langle\langle \otimes^k, A^*;1\rangle\rangle$ (where $k < n$,[20] and $\langle A^*, A\rangle \in \overline{cp}$[21]) either in D_1 or in D_2. Since $\overline{s}(A^*) = \overline{s}(A)$, $\forall_s((s \models D_1 \ \wedge \ s \models D_2) \longrightarrow s \models \langle\langle \otimes^n, A;1\rangle\rangle)$.

Case 1 and 2 show that for any infon α in $RF(D)$ that does not have an equivalent infon either in $RF(D_1)$ or in $RF(D_2)$, $\forall_s((s \models D_1 \ \wedge \ s \models D_2)$ $\longrightarrow s \models \alpha)$. Therefore, $\{D_1, D_2\} \models D$.

3.6.2 Soundness theorem

Soundness Theorem *If* $\Delta \vdash D$, *then* $\Delta \models D$.

Proof Suppose that $\Delta \vdash D$. By definition, there is a sequence of *wfd*s $< D_1, \ldots, D_n >$ such that $D_n \equiv_{cp} D$ and for each k $(1 \leq k \leq n)$ either (a) there is some D' such that $D' \in \Delta$ and $D' \equiv_{cp} D_k$, or (b) there is a set of diagrams, Δ', such that for every diagram D_i in Δ', $i < k$ and $\Delta' \rightsquigarrow D_k$. We show by induction on the length of a sequence of *wfd*s that for any diagram D obtainable from Δ, D follows from Δ.

[19]It is impossible for both A_1 and A_2 to be in the same diagram, D_1 or D_2. If they were, then $A_1 + A_2$, A_1-*and*-A_2, $A_1 - A_2$, and $A_2 - A_1$ would be in that diagram, which contradicts our assumption that A has no equivalent region to either of the given diagrams.

[20]It should be noted that $n > 1$. Suppose that $\langle\langle \otimes^1, A;1\rangle\rangle \in RF(D)$. But, we assumed that there is no equivalent representing fact about D_1 or D_2. However, the unification rule does not allow us to introduce $\otimes^1$ in a region that does not have any equivalent region. Compare clause 4 of the rule of unification of diagrams.

[21]Compare clause 4-ii of the unification rule.

(Basis case) This is when the length of the sequence is 1. That is, $D_1 \equiv_{cp} D$. Since there is no previous diagram in this sequence, there are only two cases: (1) There is some D' such that $D' \in \Delta$ and $D' \equiv_{cp} D_1$. Since $D' \in \Delta$, $\Delta \models D'$. Since $D_1 \equiv_{cp} D$ and $D' \equiv_{cp} D_1$, $D' \equiv_{cp} D$. Therefore, $\Delta \models D$. (2) $\Delta' \leadsto D_1$, where $\Delta' = \emptyset$, and $D_1 \equiv_{cp} D$. D_1 must be a diagram with a rectangle only. Hence, $RF(D_1) = \emptyset$. Trivially, $\Delta \models D_1$. Since $D_1 \equiv_{cp} D$, $\Delta \models D$.

(Inductive step) Suppose that for any *wfd* D if D has a sequence of length less than n, then $\Delta \models D$. We want to show that if *wfd* D has a sequence of length n then $\Delta \models D$. That is, $D_n \equiv_{cp} D$. If there is some D' such that $D' \in \Delta$ and $D' \equiv_{cp} D_n$, then as we proved in the basis case, $\Delta \models D$. Otherwise, it must be the case that there is a set of diagrams Δ' such that for every diagram D_i in Δ' i is less than n and $\Delta' \leadsto D_n$. By the inductive hypothesis, for any D_i, if i is less than n, $\Delta \models D_i$. That is, $\forall_{D_i \in \Delta'} \Delta \models D_i$. But, we proved in the previous subsection that if $\Delta' \leadsto D_n$, then $\Delta' \models D_n$. By the transitivity of the consequence relation, $\Delta \models D_n$. Since $D_n \equiv_{cp} D$, $\Delta \models D$.

This completes the proof of the soundness of this deductive system. □

3.7 Completeness

In this section, we raise the question of the completeness of this deductive system. Whenever *wfd* D follows from the diagrams in Δ (i.e., $\Delta \models D$), is it the case that *wfd* D is obtainable from a set Δ of *wfd*s (i.e., $\Delta \vdash D$) in this system? In the first subsection, I will introduce a new notion, *maximal diagram*, and prove that maximal diagrams have some interesting properties, which will be used crucially in the completeness proof in the second subsection.

In this work, I will limit myself to a finite set of diagrams. The method I will present in this section, however, can be extended to an infinite set of diagrams.[22]

3.7.1 Maximal diagram

Suppose that for a given diagram, say, D, we obtain a diagram that contains all the logical consequences of D. Then, most probably we

[22] Eric Hammer and Norman Danner, in their unpublished paper, "Toward a Model Theory of Venn Diagrams," used my result and method to prove the completeness of this system for an infinite set of diagrams.

could erase information from this maximal diagram and get a diagram we want, which the completeness proof aims for. Therefore, for a given diagram D, we are interested in getting a diagram, say, D^*, such that D^* contains all and only logical consequences of the given diagram D. This intuitive idea can be captured in the following form:

For given diagram D and given counterpart relation cp,
D^* is a maximal diagram of D iff $\forall_{D'}(\{D\} \models D' \longleftrightarrow D' \subseteq_{cp} D^*)$.
(I.e., D^* contains all and only diagrams that are consequences of D.)

However, this intuitive idea needs to be modified. The following example will show how we will get a more accurate definition of this new notion.

Example 39 *Let* $\langle A_1, A_2 \rangle$, $\langle A_2, A_3 \rangle$, $\langle A_3, A_4 \rangle$, $\langle A_4, A_5 \rangle \in cp$.

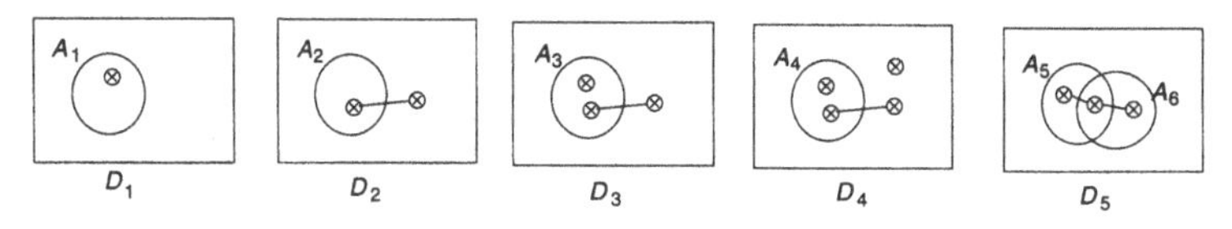

D_2 is a consequence of D_1, but D_1 does not include D_2. Therefore, we do not want to say that D_1 is a maximal diagram of D_2. However, D_3 seems to include whatever follows from D_1, but nothing else. D_4 contains D_3, but it also contains the following diagram which is not a consequence of D_1 (let $\langle A_1, A_7 \rangle \in cp$):

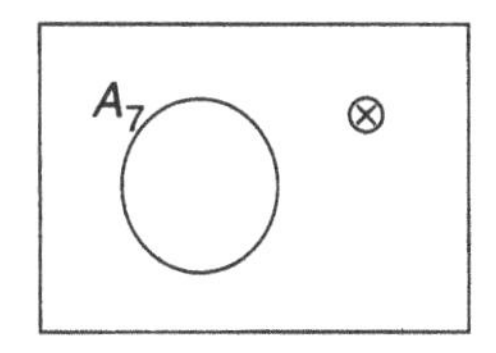

Accordingly, D_4 is not a maximal diagram of D_1. The next question is whether or not D_3 contains every diagram that is a consequence of D_1. Diagram D_5 answers this question in the negative: D_5 is a consequence of D_1, but D_3 does not include D_5.

This example shows that we need a restriction on basic regions for the definition of a maximal diagram. Otherwise, no diagram would be a maximal diagram of a given diagram, since we could always draw a new closed curve so that this new diagram is a logical consequence of the given diagram.

Definition *D^* is a maximal diagram (or maximal) of D iff D^* includes all and only diagrams:*

(i) *which are consequences of D and*
(ii) *whose basic regions are included in the basic regions of D.*

(This definition can be formalized as follows: D^* is a maximal diagram of D iff for a given counterpart relation cp, $\forall_{D'}((\{D\} \models D' \ \wedge \ BR(D') \subseteq_{cp} BR(D)) \longleftrightarrow D' \subseteq_{cp} D^*)$.)

According to this revised definition, in Example 39, D_3 is a maximal diagram of D_1. We will examine two more examples in order to understand the notion of a maximal diagram more clearly.

Example 40 *Let $\langle A_1, A_3 \rangle$, $\langle A_2, A_4 \rangle$ $\langle A_3, A_5 \rangle$, $\langle A_4, A_6 \rangle \in cp$.*

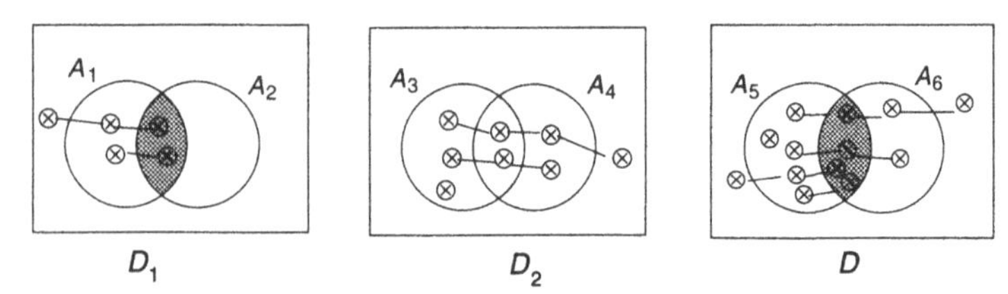

D_2 is a consequence of D_1, but D_2 is not included in D_1. However, D includes any diagram D' such that D' follows from D_1 and the basic regions of D' are included in the basic regions of D_1. Therefore, D is a maximal diagram of D_1. However, D is not a maximal diagram of D_2, since D includes the following diagram which does not follow from D_2:

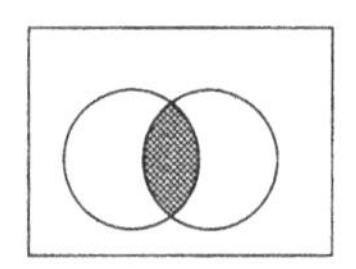

Example 41 *This example shows us how the condition for basic regions works. Let $\langle A_1, A_2 \rangle$, $\langle A_1, A_3 \rangle \in cp$.*

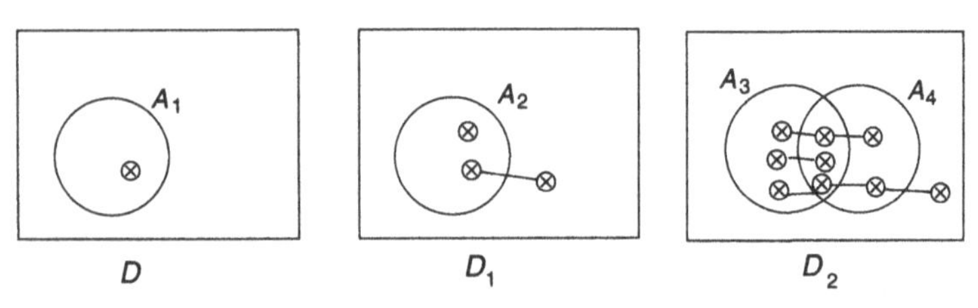

In this case, D_1 is a maximal of D, while D_2 is not a maximal of D since there is a region, A_4, in D_2 that does not have any equivalent region in D.

Now, we want to prove the existence of a maximal diagram for any given diagram, which is called the maximal representation theorem. This theorem says that for any diagram D Venn-I allows us to obtain its maximal diagram.

Theorem 11 (Maximal Representation Theorem) *For every diagram D, there is a maximal diagram D^* of D such that $\{D\} \vdash D^*$.*

Proof First, I will show how to construct a maximal diagram of a given diagram, D. And I will prove that this diagram is a maximal of D.

If D has a region with both a shading and an x-sequence, by the rule of conflicting information we can get any diagram we want. Trivially, $\{D\} \vdash D^*$. Otherwise, for a given diagram D, the following is the procedure to get D^*.

(Step 1) Is there any region y and z such that

1. $\langle\langle \otimes^n, y; 1\rangle\rangle \in RF(D)$,
2. $\langle\langle \text{Shading}, z, 1\rangle\rangle \in RF(D)$, and
3. there is an overlapping region between y and z?[23]

If not, copy D and name it D'. Go to step 2.
If so (suppose there are m cases that satisfy the previous conditions), then apply the rule of the erasure of part of an x-sequence m times in the following way:

$\{D\} \leadsto D_1$ (erase the part of an x-sequence in *y-and-z* by the rule of the erasure of part of an x-sequence),

$$\{D_1\} \leadsto D_2,$$

$$\vdots$$

$$\{D_{m-1}\} \leadsto D_m.$$

Apply the rule of unification in the following way:

$$\{D, D_m\} \leadsto D'.$$

The following example shows how this step works:

[23]We know that there is only a partially overlapping region between y and z. Otherwise, D has a region with both a shading and an x-sequence, which contradicts our assumption.

Example 42 *When D is given, this step tells us to transform D to D′ in the following way:*

(1) $\{D\} \rightsquigarrow D_1$ (by the rule of erasure of part of an x-sequence).
(2) $\{D_1\} \rightsquigarrow D_2$ (by the rule of erasure of part of an x-sequence).
(3) $\{D, D_2\} \rightsquigarrow D'$ (by the rule of unification of diagrams).

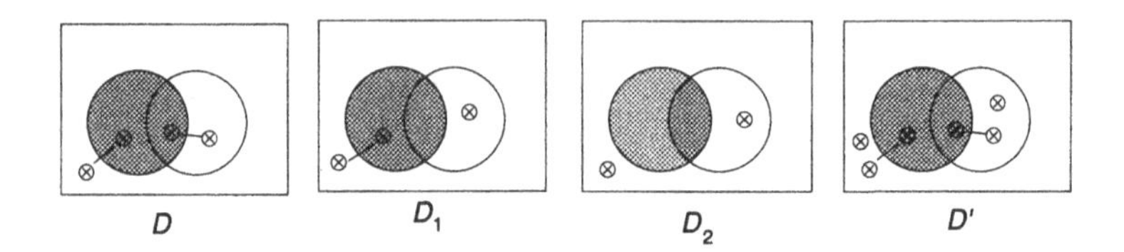

(Step 2) In D', is there any x-sequence?
If not, copy D' and name it D^*.
If so (suppose that there are k x-sequences), apply the rule of spreading x's in the following way:

Since there are k x-sequences,

$$\langle\!\langle \otimes^{i_1}, A_1; 1\rangle\!\rangle \in RF(D'),$$
$$\vdots$$
$$\langle\!\langle \otimes^{i_k}, A_k; 1\rangle\!\rangle \in RF(D').$$

Let set $M(A_j)$ (when $1 \leq j \leq k$) be the set of minimal regions of region $U - A_j$ (where U is the rectangle). We want to spread x's to each member of set $\mathscr{P}(M(A_j)) - \{\emptyset\}$ by applying the rule of spreading x's. Therefore, for each x-sequence, we apply this rule $|\mathscr{P}(M(A_j)) - \{\emptyset\}|$ times. That is,

$$\{D'\} \rightsquigarrow D_1^1,\ \{D_1^1\} \rightsquigarrow \ldots \rightsquigarrow D^1_{|\mathscr{P}(M(A_1))-\{\emptyset\}|}$$
$$\vdots$$
$$\{D'\} \rightsquigarrow D_1^k,\ \{D_1^k\} \rightsquigarrow \ldots \rightsquigarrow D^k_{|\mathscr{P}(M(A_k))-\{\emptyset\}|}$$

Unify the diagrams, $D', \ldots, D^1_{|\mathscr{P}(M(A_1))-\{\emptyset\}|}, \ldots, D^k_{|\mathscr{P}(M(A_k))-\{\emptyset\}|}$ by the rule of unification.[24] Call the unified diagram from all these diagrams D^*.

Let us look at the following example to see how step 2 works:

Example 43 *Given D', we get D^* in the following way:*

(1) $\{D'\} \vdash D_1^1$ (by the rule of spreading x's).

[24]We unify two at a time. For example, $\{D', D_1^1\} \rightsquigarrow D_1', \{D_1', D_2^1\} \rightsquigarrow D_2', \ldots$

(2) $\{D'\} \vdash D_1^2$ (by the rule of spreading x's).
(3) $\{D', D_1^1\} \vdash D_1^*$ (by the rule of unification of diagrams).
(4) $\{D_1^*, D_1^2\} \vdash D^*$ (by the rule of unification of diagrams).

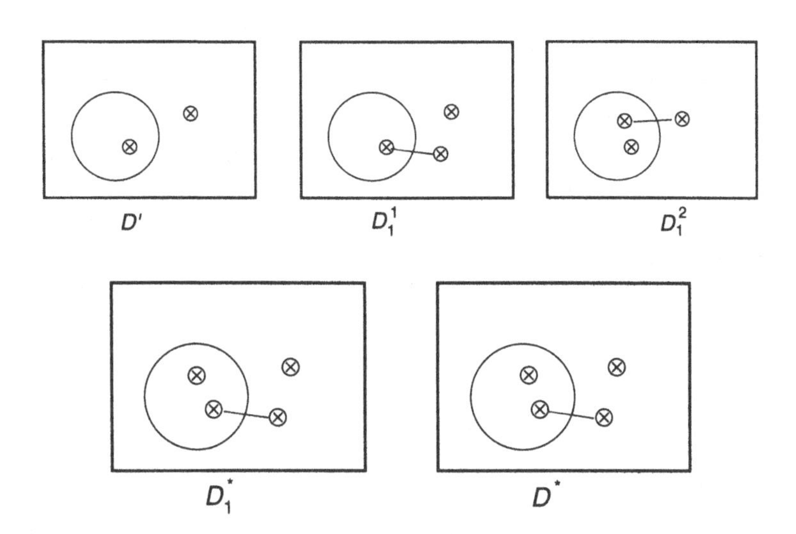

Claim *D^* is a maximal of D (i.e., $\forall_{D'}((BR(D') \subseteq_{cp} BR(D) \wedge \{D\} \models D') \longleftrightarrow D' \subseteq_{cp} D^*)$).*

Proof ($\longrightarrow$) Let D' be such that $BR(D') \subseteq_{cp} BR(D)$ and $\{D\} \models D'$. (We show that $D' \subseteq_{cp} D^*$.)

(i) Suppose that basic region $A_1 \in BR(D')$. Since $BR(D') \subseteq_{cp} BR(D)$ and $BR(D) \equiv_{cp} BR(D^*)$, there is basic region A_2 such that $\langle A_1, A_2 \rangle \in cp$ and $A_2 \in BR(D^*)$.

(ii) Suppose that $A_1 \in RShading(D')$. (We must show that there is a region A_2 such that $\langle A_1, A_2 \rangle \in \overline{cp}$ and $A_2 \in RShading(D^*)$.) Suppose that there is no region A_2 in D^* such that $\langle A_1, A_2 \rangle \in \overline{cp}$ and $A_2 \in RShading(D^*)$. Since $BR(D') \subseteq_{cp} BR(D)$ and D^* was constructed in such a way that $BR(D) \equiv_{cp} BR(D^*)$, we know that D^* has region A_2 such that $\langle A_1, A_2 \rangle \in \overline{cp}$. This implies that $\langle\langle \text{Shading}, A_2; 1 \rangle\rangle \notin RF(D^*)$, where $\langle A_1, A_2 \rangle \in \overline{cp}$. Since we did not erase any shading from D, $\langle\langle \text{Shading}, A_3; 1 \rangle\rangle \notin RF(D)$, where $\langle A_1, A_3 \rangle \in \overline{cp}$. (We know that there is a region A_3 in D that is equivalent to A_1 since $BR(D') \subseteq_{cp} BR(D)$.) So far, we have the following:

$$\langle\langle \text{Shading}, A_1, 1 \rangle\rangle \in RF(D'), \tag{1}$$

$$\langle\langle \text{Shading}, A_3, 1\rangle\rangle \notin RF(D), \tag{2}$$

$$\langle A_1, A_3\rangle \in \overline{cp}. \tag{3}$$

Let set assignment s_1 be such that $s_1 \models D$ and $\overline{s_1}(A_3) \neq \emptyset$. Since $\overline{s_1}(A_3) = \overline{s_1}(A_1)$, by equation (1) $s_1 \not\models D'$. That is, $\exists_s(s \models D \wedge s \not\models D')$, which contradicts the assumption that $\{D\} \models D'$. Therefore, there is a region A_2 such that $\langle A_1, A_2\rangle \in \overline{cp}$ and $A_2 \in RShading(D^*)$.

(iii) Suppose that $A_1 \in R\otimes^n(D')$. Suppose that there is no region A_2 in D^* such that $\langle A_1, A_2\rangle \in \overline{cp}$ and $A_2 \in R\otimes^n(D^*)$. By the construction of D^* and by the assumption that $BR(D') \subseteq_{cp} BR(D)$, we know that D^* has region A_2 such that $\langle A_1, A_2\rangle \in \overline{cp}$. This implies that $\langle\langle \otimes^n, A_2; 1\rangle\rangle \notin RF(D^*)$, where $\langle A_1, A_2\rangle \in \overline{cp}$. Since $BR(D) \equiv_{cp} BR(D^*)$, we know that there is a region in D, say, A_3, such that $\langle A_2, A_3\rangle \in \overline{cp}$. By the process of constructing D^* from D, we know the following about D:

$$\langle\langle \otimes^m, A_3, 1\rangle\rangle \notin RF(D), \tag{4}$$

since we unified D into D^* and $\langle\langle \otimes^n, A_2; 1\rangle\rangle \notin RF(D^*)$.

$$\langle\langle \otimes^k, A_i, 1\rangle\rangle \notin RF(D), \tag{5}$$

for any A_i such that A_i is a part of A_3. (Otherwise, by step 2 of our construction of D^* – the step where the rule of spreading x's is applied – we would get $\langle\langle \otimes^n, A_2; 1\rangle\rangle \in RF(D^*)$.)

If there is a region, say, A, in D such that A_3 is a part of A and $\langle\langle \otimes^l, A; 1\rangle\rangle \in RF(D)$, then

$$\langle\langle \text{Shading}, A - A_3; 1\rangle\rangle \notin RF(D). \tag{6}$$

(Otherwise, by step 1 of our construction of D^* – the step where the rule of erasure of part of an x-sequence is applied – we would get $\langle\langle \otimes^n, A_2; 1\rangle\rangle \in RF(D^*)$.)

By equations (4), (5), and (6), we know that there are four possible cases for region A_3:

(Case 1) A_3 is shaded.

(Case 2) Part of A_3 (i.e., some region which is a part of A_3) is shaded.

(Case 3) A_3 does not have any shading or any x-sequence.

(Case 4) A partial x-sequence is in A_3. In this case, by equation (6) we know that there is a region A such that $\langle\langle \otimes^l, A; 1\rangle\rangle \in RF(D)$ and $\langle\langle \text{Shading}, A - A_3; 1\rangle\rangle \notin RF(D)$.

Among the set assignments that satisfy D, pick up a set assignment, say, s_1, such that $\overline{s_1}(A_3) = \emptyset$. (How do we know that there is a set assignment such that it satisfies D and at the same time it assigns the empty set to A_3? We just exhausted the possible cases for A_3. In case 1, if $s \models D$, then $\overline{s}(A_3) = \emptyset$. In cases 2, 3, and 4, if $s \models D$, then s can assign either the empty set or a nonempty set to A_3.) Since $\overline{s_1}(A_3) = \overline{s_1}(A_1)$, by the assumption that $A_1 \in R\otimes^n(D')$, $s_1 \not\models D'$. That is, $\exists_s(s \models D \;\wedge\; s \not\models D')$, which contradicts the assumption that $\{D\} \models D'$. Therefore, there is a region A_2 such that $\langle A_1, A_2\rangle \in \overline{cp}$ and $A_2 \in R\otimes^n(D^*)$.

By (i), (ii), (iii), and the definition of a subdiagram,[25] we know that $D' \subseteq_{cp} D^*$.

($\longleftarrow$) Let D' be such that $D' \subseteq_{cp} D^*$. Accordingly, $BR(D') \subseteq_{cp} BR(D^*)$. Since we did not add any new closed curve, $BR(D^*) \equiv_{cp} BR(D)$. Therefore, $BR(D') \subseteq_{cp} BR(D)$. Since $\{D\} \vdash D^*$, by soundness $\{D\} \models D^*$. Since $D' \subseteq_{cp} D^*$, by Theorem 9, $\{D^*\} \models D'$. Hence, by transitivity, $\{D\} \models D'$. □

Let us extend Theorem 11 to a nonempty finite set[26] diagrams to get the following corollary:

Corollary 11 *Let Δ be a nonempty set of diagrams and D be the unification of the diagrams in Δ. Then, there is a diagram D^* such that D^* is a maximal of D and $\Delta \vdash D^*$.*

Proof Obtain diagram D from the diagrams of Δ by the rule of unification: $\Delta \vdash D$. By the maximal representation theorem, we know that we can obtain D^* from D such that D^* is a maximal of D. That is, $\Delta \vdash D^*$. □

Definition *We call the diagram D^* in Corollary 11 a maximal diagram of Δ.*

Now, we need to prove the uniqueness of a maximal diagram for a given diagram.

Theorem 12 (Uniqueness of a Maximal Diagram) *Let D, D_1, and D_2 be wfds and cp be a given counterpart relation for these diagrams. Let D be*

[25] Refer to §3.4.1.

[26] As said at the beginning of this section, I will deal with a finite set of diagrams only in this work. Accordingly, from now on, I will refer to a finite set of diagrams only as a set of diagrams, for example, Δ, Δ'.

a consistent diagram.[27] *If D_1 and D_2 are maximal diagrams of D, then $D_1 \equiv_{cp} D_2$.*

Proof Let $Seq(D_1) = \langle BR(D_1),\ RShading(D_1),\ R\otimes^n(D_1)\rangle$ and $Seq(D_2) = \langle BR(D_2),\ RShading(D_2),\ R\otimes^n(D_2)\rangle$. Since D_1 and D_2 are maximal diagrams of D, by the definition of a maximal diagram $BR(D_1) \equiv_{cp} BR(D)$ and $BR(D_2) \equiv_{cp} BR(D)$. Hence, $BR(D_1) \equiv_{cp} BR(D_2)$. Suppose that $D_1 \not\equiv_{cp} D_2$. Then, it is the case either that $RShading(D_1) \not\equiv_{cp} RShading(D_2)$ or that $R\otimes^n(D_1) \not\equiv_{cp} R\otimes^n(D_2)$. Suppose $RShading(D_1) \not\equiv_{cp} RShading(D_2)$. Hence, there is a region a such that $a \in RShading(D_1)$ but $\overline{cp}(a) \notin RShading(D_2)$. That is, $\langle\langle$Shading, $a;1\rangle\rangle \in RF(D_1)$, but $\langle\langle$Shading, $\overline{cp}(a);1\rangle\rangle \notin RF(D_2)$ (where $\overline{cp}(a) \in RG(D_2)$). Since $\langle\langle$Shading, $a;1\rangle\rangle \in RF(D_1)$, this representing fact is a consequence of D. (Recall that a maximal diagram contains all and *only* information that is a consequence of a given diagram.) Then, this representing fact should be included in D_2, since D_2 is a maximal of D by our assumption. Therefore, this is a contradiction. Other cases are similar to this. □

The following is an interesting relation between a given diagram and its maximal diagram:

Theorem 13 *Let D^* be a maximal of D. Then, D and D^* are logical consequences of each other.* (I.e., $\{D^*\} \models D$ *and* $\{D\} \models D^*$.)

Proof Since D $\subseteq_{cp} D^*$, by Theorem 9, $\{D^*\} \models D$. By the Maximal Representation Theorem and the Soundness Theorem, $\{D\} \models D^*$. □

Corollary 13 *Let D^* be a maximal of D. Then, $\forall_{D'}(\{D\} \models D' \longleftrightarrow \{D^*\} \models D')$.*

Using the previous definition for a maximal diagram of set Δ and Theorem 13, we prove the following feature which will be used in the completeness proof:

Theorem 14 *Let D^* be a maximal of Δ. Then, every diagram of Δ is a consequence of D^*. (I.e., $\forall_{D\in\Delta}(\{D^*\} \models D)$.)*

[27]**Definition** *Wfd D is consistent iff D does not have a region with both a shading and an x-sequence.*

Proof Let D' be the diagram we obtain from Δ by the unification rule. Accordingly, every representing fact of each diagram of Δ is included in the representing facts of D'. Hence, $\forall_{D\in\Delta}(\{D'\} \models D)$. Since D^* is a maximal of D', by Theorem 13 $\{D^*\} \models D'$. Therefore, $\forall_{D\in\Delta}(\{D^*\} \models D)$. □

Maximal diagrams of given diagrams, regardless of these given diagrams, share significant properties that we will need for the completeness proof. Before we prove these properties, let us have the following definition for convenience.

Definition *D^* is a maximal diagram (or maximal) iff there is a diagram D such that D^* is a maximal of D.*

The next two theorems prove some facts about maximal diagrams, which will be used in the completeness proof.

Theorem 15 *If D^* is a maximal, then D^* is a maximal of itself (i.e., $\forall_{D'}((\{D^*\} \models D' \wedge BR(D') \subseteq_{cp} BR(D^*)) \longleftrightarrow D' \subseteq_{cp} D^*)$, where cp is a given counterpart relation.)*

Proof Suppose D^* is maximal.

($\longrightarrow$) Let D' be such that $\{D^*\} \models D'$ and $BR(D') \subseteq_{cp} BR(D^*)$. (We show that $D' \subseteq_{cp} D^*$.) Let D be a diagram such that D^* is a maximal of D and cp be a given counterpart relation. By Theorem 13,

$$\{D\} \models D^*. \tag{1}$$

We have the assumption that

$$\{D^*\} \models D'. \tag{2}$$

By equations (1) and (2),

$$\{D\} \models D'. \tag{3}$$

Since D^* is a maximal of D,

$$BR(D^*) \equiv_{cp} BR(D). \tag{4}$$

We have the assumption that

$$BR(D') \subseteq_{cp} BR(D^*). \tag{5}$$

By equations (4) and (5),

$$BR(D') \subseteq_{cp} BR(D). \tag{6}$$

By equations (3) and (6) and the definition of a maximal diagram (i.e., $\forall_{D'}((\{D\} \models D' \ \wedge \ BR(D') \subseteq_{cp} BR(D)) \longleftrightarrow D' \subseteq_{cp} D^*))$, we know that

$$D' \subseteq_{cp} D^*. \tag{7}$$

($\longleftarrow$) Let D' be such that $D' \subseteq_{cp} D^*$. (We show that $\{D^*\} \models D'$ and $BR(D') \subseteq_{cp} BR(D^*)$.) By Theorem 9, $\{D^*\} \models D'$, and obviously $BR(D') \subseteq_{cp} BR(D^*)$ (since $D' \subseteq_{cp} D^*$).

This completes the proof. □

Since a maximal diagram is a maximal of itself, along with the uniqueness proof of a maximal diagram, we know that if a given diagram, say, D, is maximal, then $D^* \equiv_{cp} D$ (where D^* is a maximal of D).

We discussed earlier why we needed a restriction on basic regions for the definition of a maximal diagram. However, for a given diagram we can always draw a maximal diagram that includes that given diagram, since we have a rule of introduction of a basic region. Importantly, this maximal diagram might not always be a maximal diagram of the given diagram. This will be used in our completeness proof.

Theorem 16 *Let D be a wfd and D' be a consequence of D. Then, there is a maximal diagram D_0^* such that $\{D\} \vdash D_0^*$ and D_0^* includes D'. Notice that D_0^* does not have to be a maximal diagram of D, but of some other diagram. (I.e., $\forall_{D'}(\{D\} \models D' \longrightarrow \exists_{D_0^*}(D_0^*$ is maximal $\wedge \{D\} \vdash D_0^* \ \wedge \ D' \subseteq_{cp} D_0^*))$, where cp is a given counterpart relation among D, D', and D_0^*.)*

Proof If $BR(D') \subseteq_{cp} BR(D)$, by Theorem 15 $D_0^* = D^*$ (where D^* is a maximal of D). Otherwise, keep applying the rule of introduction of a closed curve to D to get D_0 (where $BR(D') \subseteq_{cp} BR(D_0)$). That is,

$$\{D\} \vdash D_0, \quad \text{where } BR(D') \subseteq_{cp} BR(D_0). \tag{1}$$

Let D_0^* be a maximal diagram of D_0. Then,

$$\{D_0\} \vdash D_0^* \quad \text{(by the maximal representation theorem)}, \tag{2}$$

$$\{D_0^*\} \models D_0 \quad \text{(by Theorem 13)}, \tag{3}$$

$$\{D_0\} \vdash D \quad \text{(by erasing closed curves)}. \tag{4}$$

By equation (4) and soundness,

$$\{D_0\} \models D. \tag{5}$$

By equations (3) and (5) and the assumption that $\{D\} \models D'$,

$$\{D_0^*\} \models D'. \tag{6}$$

Since D_0^* is a maximal of D_0, $BR(D_0) \equiv_{cp} BR(D_0^*)$. By equation (1),

$$BR(D') \subseteq_{cp} BR(D_0^*). \tag{7}$$

By equations (6) and (7) Theorem 15,

$$D' \subseteq_{cp} D_0^*. \tag{8}$$

By equations (1) and (2),

$$\{D\} \vdash D_0^*. \tag{9}$$

Equations (8) and (9) complete the proof. □

We want to use maximal diagrams to prove the completeness of this system. However, as seen before, if the empty set is given to us, we cannot obtain a maximal diagram of that empty set. Therefore, before the completeness proof, we want to note a property of the diagrams that follow from the empty set.

Lemma (tautology lemma) Let $\Delta = \emptyset$ *and* D^0 *be a wfd consisting only of a rectangle. Then,* $\forall_D(\Delta \models D \longleftrightarrow \{D^0\} \models D)$.

Proof ($\longrightarrow$) Let D be such that $\emptyset \models D$. Accordingly, $\forall_s(s \models D)$. Suppose that $\{D^0\} \not\models D$. That is, $\exists_s(s \models D^0 \ \wedge \ s \not\models D)$. This is a contradiction. ($\longleftarrow$) Let D be such that $\{D^0\} \models D$. So, $\forall_s(s \models D^0 \longrightarrow s \models D)$. But, since there is no representing fact in D^0, $\forall_s(s \models D^0)$. Therefore, $\forall_s(s \models D)$. Accordingly, $\emptyset \models D$. □

3.7.2 Completeness theorem

Completeness Theorem *If* $\Delta \models D$, *then* $\Delta \vdash D$.

Proof Suppose that $\Delta \models D$.
(Case 1) $\Delta = \emptyset$. Let D^0 be a *wfd* consisting only of a rectangle. By applying the rule of introduction of a basic region, we get D^0. Hence, $\Delta \vdash D^0$. Let D^* be a maximal of D^0. In this case, obviously $D^0 \equiv_{cp} D^*$. That is, $\{D^0\} \vdash D^*$. By transitivity, $\Delta \vdash D^*$. By the tautology lemma, $\{D^0\} \models D$. Since D^* is a maximal of D^0, $\{D^*\} \models D^0$. By transitivity, $\{D^*\} \models D$.
(Case 2) $\Delta \neq \emptyset$. Let D^* be a maximal diagram of Δ. By Corollary 11, we get D^* from Δ. That is, $\Delta \vdash D^*$. By Theorem 14, $\{D^*\} \models D'$ for every $D' \in \Delta$. Since $\Delta \models D$ by the assumption, $\{D^*\} \models D$.

In either case, we get the following:

$$\Delta \vdash D^*. \tag{1}$$

$$\{D^*\} \models D. \tag{2}$$

By equation (2) and Theorem 16, we know that there is a maximal diagram D_0^* such that

$$\{D^*\} \vdash D_0^*, \tag{3}$$

$$D \subseteq_{cp} D_0^*. \tag{4}$$

By equation (4) and Theorem 9,

$$\{D_0^*\} \vdash D. \tag{5}$$

By equations (1), (3), and (5),

$$\Delta \vdash D. \tag{6}$$

This completes the proof of the completeness of this deductive system. □

4
Venn-II

Venn-I, which I proved to be sound and complete in the previous chapter, has been used to test the validity of a specific form of syllogism, that is, the categorical syllogism. A categorical syllogism consists of two premises and a conclusion, all of which are categorical sentences only. Can Venn-I test the validity of the following syllogism, which is not a categorical syllogism even though all the components of the premises and the conclusion are categorical sentences?

> If no unicorn is red, then every unicorn is red.
> All red things are visible.
>
> ---
> Every unicorn is visible or some unicorns are visible.

Venn-I cannot represent the information that the first premise and the conclusion convey. Therefore, the method of diagrams in Venn-I is not adequate for testing this syllogism. In his book, *Methods of Logic*, Quine suggests the following seemingly complicated example as an argument where "the unaided method of diagrams [Venn-I] bogs down":[1]

Premises: If all applicants who received the second announcement are of the class of '00, then some applicants did not receive the second announcement.

Either all applicants received the second announcement or all applicants are of the class of '00.

Conclusion: If all applicants of the class of '00 received the second announcement then some applicants not of the class of '00 received the second announcement.[2]

Quine calls this example an argument with an *admixture of truth function* in that some sentences in this argument consist of both categorical

[1]Quine [20], p. 95
[2]Quine [20], p. 95

sentences and truth functional connectives, that is, "if ... then" or "or." While Venn-I can represent the information expressed in categorical sentences, it fails in representing the information conveyed in sentences with these truth-functional connectives. On the other hand, a sentential language can handle truth-functional connectives, but not categorical sentences appropriately. Quine's remark suggests that another language should be introduced to combine these two methods:

> Diagrams [Venn-I] are suited to handling the components 'All F who are G are H', 'Some F are not G', etc., and the methods of [sentential logic] are suited to '→', and '∨'; but just how may we splice the two techniques in order to handle a combined inference of the above kind?[3]

A first-order language with quantifiers and monadic predicates is introduced after this passage. The two syllogisms given previously can be translated into this first-order language in the following way:

$$\forall x(Ux \to \neg Rx) \to \forall x(Ux \to Rx).$$
$$\forall x(Rx \to Vx).$$

$$\forall x(Ux \to Vx) \vee \exists x(Ux \wedge Vx).$$

$$\forall x((Fx \wedge Gx) \to Hx) \to \exists x(Fx \wedge \neg Gx).$$
$$\forall x(Fx \to Gx) \vee \forall x(Fx \to Hx).$$

$$\forall x((Fx \wedge Hx) \to Gx) \to \exists x((Fx \wedge \neg Hx) \wedge Gx)).$$

Obviously, Venn-I is more limited than this first-order language in the degree of expressiveness. In this chapter, I extend Venn-I to be equivalent to a monadic first-order language. I call this new system Venn-II, and will present the syntax and the semantics of this extended system as the syntax and the semantics of Venn-I were in the previous chapter.

The syntax and the semantics of Venn-II will be extended in the following way. The two Venn diagram systems, Venn-I and Venn-II, have the same objects as primitives. In order to increase the expressiveness, a new rule of well-formedness is added for Venn-II, which will be given in the second section of this chapter. This syntactic change in meaningful units of a system requires a change in the semantics. While there will be

[3]Quine [20], p. 95

no change in the way set assignments and representing facts are defined, a new inductive clause for the set of *wfd*s requires that the satisfaction relation between a diagram and a set assignment be defined inductively. The definition of satisfaction for Venn-I will be kept as a basis clause and a new inductive clause will be introduced. This semantic change will be spelled out in the third section. The new formation rule also requires one transformation rule of Venn-I to be modified, one rule to be extended, and four new transformation rules to be introduced. We will stipulate these changes in the fifth section. Again, I will prove that this new extended system, Venn-II, is sound and complete in §6 and §7, respectively.

4.1 Preliminary remarks

Before constructing the language of Venn-I, we discussed the features we desired to incorporate into this system. Now, we are going to extend Venn-I to increase the degree of expressiveness of a Venn-diagram representation system. Accordingly, all the features incorporated into Venn-I will be kept in this new system. Let us consider the features we want to add for this extended system. As seen before, Venn-I cannot represent the information that the following sentences convey:

> If every unicorn is red, *then* some unicorn is visible.
> Every unicorn is red *or* some unicorn is visible.

These sentences consist of categorical sentences and truth-functional connectives. Let us define a set of extended categorical sentences, call it set EC, to be the smallest set satisfying the following:

(i) Every categorical sentence is in EC.
(ii) If α and β are in EC, then so is $(\alpha \wedge \beta)$.
(iii) If α is in EC, then so in $(\neg\alpha)$.
(iv) If α and β are in EC, then so is $(\alpha \vee \beta)$.
(v) If α and β are in EC, then so is $(\alpha \rightarrow \beta)$.

We want an extended system to be able to represent any information conveyed by any member of set EC. That is, we want the following to be true:

(1) Every categorical sentence is expressible in Venn-II.
(2) If α and β are expressible in Venn-II, so is $(\alpha \wedge \beta)$.
(3) If α is expressible in Venn-II, then so is $(\neg\alpha)$.
(4) If α and β are expressible in Venn-II, so is $(\alpha \vee \beta)$.
(5) If α and β are expressible in Venn-II, so is $(\alpha \rightarrow \beta)$.

Categorical sentences can be mirrored in Venn-I. The rule of unification of diagrams allows us to unify two pieces of information represented in diagrams. Therefore, the following is true of Venn-I:

(1) Every categorical sentence is expressible in Venn-I.
(2) If α and β are expressible in Venn-I, so is $(\alpha \wedge \beta)$.

Venn-I also has a way to express some disjunctive information. Connected ⊗'s, that is, an x-sequence, can represent disjunctive information among existential statements. For example, "some unicorn is visible *or* some unicorn is red" is expressible in Venn-I. However, this syntactic device is too limited to cover all possible combinations of disjunctive information among categorical sentences. We extend Venn-I in such a way that clauses 3, 4, and 5 hold true as well as 1 and 2. If we invent a syntactic mechanism to make clause 3 true,[4] then clauses 4 and 5 will be true by the fact that $\{\neg, \wedge\}$ is truth-functionally complete. However, I find it easier to introduce a way to handle the information conveyed by a disjunction, rather than that conveyed by a negation. Now the question is whether introducing a way to represent disjunctive information is sufficient to represent any information conveyed by a member of set *EC*. We need to prove the following:[5]

Theorem 17 *Suppose the following:*

(1) *Every categorical sentence is expressible in Venn-II.*
(2) *If α and β are expressible in Venn-II, so is $(\alpha \wedge \beta)$.*
(3) *If α and β are expressible in Venn-II, so is $(\alpha \vee \beta)$.*

Then, if α is expressible in Venn-II then so is $(\neg\alpha)$.

Proof We know that all the negation signs of $(\neg\alpha)$ can be pushed inside until they apply to categorical sentences. However, the negation of a categorical sentence yields another categorical sentence.[6] Therefore, by (2) and (3), we know that $(\neg\alpha)$ is expressible in Venn-II. □

[4]For example, in his existential graph, Peirce encircles a proposition to negate it. The following is the negation of the proposition "some unicorn is red."

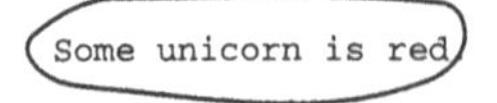

[5]If this theorem is true, then clause 5 will be true by the fact that either $\{\neg, \wedge\}$ or $\{\neg, \vee\}$ is truth-functionally complete.

[6]"It is not the case that every unicorn is red" conveys the same information as "Some unicorns are not red" and so on.

Therefore, we will introduce a new syntactic device in order to represent disjunctive information. Let us recall Peirce's suggestions for disjunctive information, which were discussed in the second chapter. First, Peirce introduced a line that connects syntactic objects. However, to adopt this we would have to sacrifice some visual power of the diagram. Peirce's other suggestion is to put Venn diagrams into rectangles and to interpret the Venn diagrams in each rectangle as disjunct. In this way, each of the Venn diagram does not have to have a confusing line between o's or x's. However, in Venn-I we introduced a rectangle to represent a background set. Therefore, we will not adopt Peirce's method directly, but will connect diagrams by a line. In this way, we do not have to introduce any new syntactic object. Recall that in Venn-I we have a line in an x-sequence in order to represent disjunctive information. Since we do not have a connected diagram as a meaningful unit in Venn-I, a new clause is needed in the well-formedness of diagrams. Accordingly, there will be a slight change in the definitions of counterpart relations. All of these will be the topics of the following section.

4.2 Well-formed diagrams

4.2.1 Primitive objects

As discussed in the preliminary remarks, we do not need any new primitive object. Venn-II has the same set of primitive objects as Venn-I as follows:

Diagrammatic Objects	Name
⬯	closed curve
▭	rectangle
●	shading
⊗	x
—	line

4.2.2 Well-formed diagrams

The set of well-formed diagrams, say, $\mathscr{D}$, is the smallest set satisfying the following rules:

(1) Any unique rectangle drawn in a plane is in set $\mathscr{D}$.

(2) If D is in the set $\mathscr{D}$, and if D' results by adding a closed curve interior to the rectangle of D by the partial-overlapping rule (described subsequently) and by the avoid-⊗ rule (described subsequently), then D' is in set $\mathscr{D}$.

Partial-overlapping rule: A new closed curve should overlap only *once* and only *part* of *every* existent *nonrectangle* minimal region.

Avoid-⊗ rule: A new closed curve should avoid every ⊗ of any existing x-sequence.

(3) If D is in the set $\mathscr{D}$, and if D' results by shading some entire region of D, then D' is in set $\mathscr{D}$.

(4) If D is in the set $\mathscr{D}$, and if D' results by adding an ⊗ to a minimal region of D, then D' is in set $\mathscr{D}$.

(5) If D is in the set $\mathscr{D}$, and if D' results by connecting existing ⊗'s by lines (where each ⊗ is in a different region), then D' is in set $\mathscr{D}$.

(6) If D_1 and D_2 are in the set $\mathscr{D}$, and if D' results by connecting these two diagrams by a straight line, then D' is in set $\mathscr{D}$. (We say $D' = D_1 – D_2$.)

Rule 6 is a new formation rule introduced in Venn-II in order to represent disjunctive information. This rule tells us how to connect diagrams. If a diagram has one rectangle, we call it an *atomic* diagram. If a diagram has more than one rectangle (that is, if the diagram consists of more than one atomic diagram), we call it a *compound* diagram. We call each atomic diagram of a diagram, say, D, a constituent of D. If a diagram is atomic, it has only one constituent, itself. If a diagram is compound, it has more than one constituent.

Example 44 *The following three diagrams are atomic well-formed diagrams.*

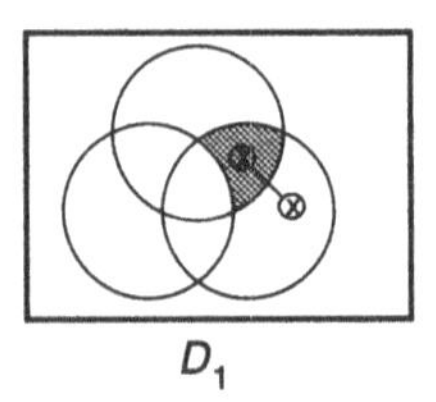

D_1

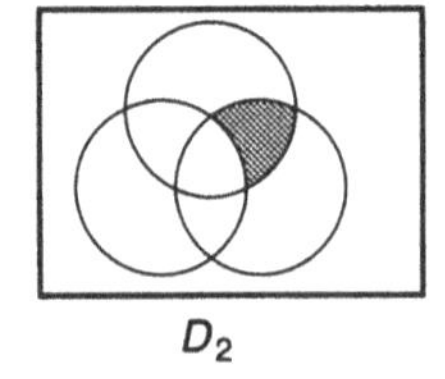

D_2

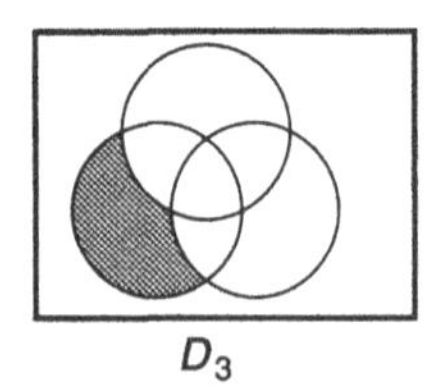

D_3

Example 45 *We connect D_1 and D_2 of Example 44 by a straight line to get D (the diagram on the left as follows). We say that $D = D_1$–D_2 and D has two constituents, D_1 and D_2. Again, we connect D and D_3 by a line to get D'. We can say either $D' = D$–D_3 or $D' = D_1$–D_2–D_3. D_1, D_2, and D_3 are the constituents of D'. The sixth rule of the well-formedness conditions tells us these two diagrams are well formed. Both of them are compound wfds.*

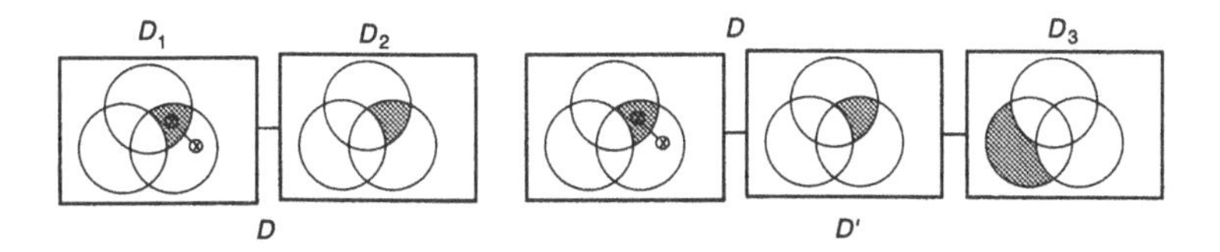

4.2.3 Counterpart relations

In the case of Venn-I, we do not allow a counterpart relation among the basic regions of the same diagram. However, in the case of Venn-II, we do not want to rule out the possibility that a counterpart relation holds among some basic regions of the same diagram if the diagram is a compound one. Still, if a user knows that set A and set B are the same set, then he would not draw two closed curves for one and the same set within one atomic diagram. Therefore, the definition of a counterpart relation is slightly modified in the following way:

Given diagrams $D_1, \ldots, D_n$, let a counterpart relation (let us call it set cp) be an equivalence relation on the set of basic regions of $D_1, \ldots, D_n$, satisfying the following:

(1) If $\langle A, B \rangle \in cp$, then both A and B are either closed curves or rectangles.
(2) If $\langle A, B \rangle \in cp$, then either A is identical to B or A and B are in different atomic diagrams.

4.3 Semantics

The definition and notation of set assignments will be kept in Venn-II as they are in Venn-I.[7] We will also keep information-theoretic terminology for the notation of a representing fact. However, in this extended system,

[7] Refer to §3.3.1.

the set of the representing facts of a diagram is not as simple as in Venn-I, since a connected diagram represents disjunctive information. We cannot simply make a union out of two different kinds of infons, that is, $\langle\langle \text{Shading}, x; 1\rangle\rangle$ and $\langle\langle \otimes^n, y; 1\rangle\rangle$, as we did in Venn-I.[8]

If we keep the same definition of satisfaction of a diagram by a set assignment without restricting the definition of the set of representing facts of a diagram, we will have to admit that the following two diagrams are satisfied by the same set assignments:

Example 46 *Let* $\langle A, B\rangle \in cp$ *and* $\langle B, C\rangle \in cp$.

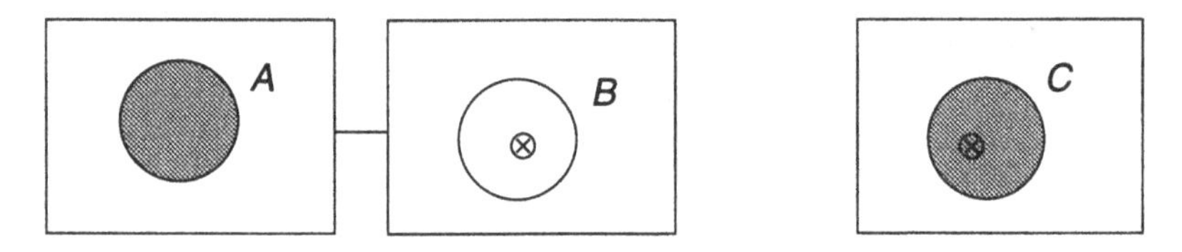

In order to prevent this undesirable result, we want to define the satisfaction relation differently. In §4.2.2, the set of well-formed diagrams was extended in this system, by introducing one new inductive rule. This new rule brought in a distinction between atomic diagrams and compound diagrams. By using this new distinction, let us define the satisfaction relation between set assignment s and diagram D inductively, as follows:

Set assignment s satisfies *wfd* D, $s \models D$, iff

(1) If D is atomic, then $\forall_{\alpha \in RF(D)}(s \models \alpha)$.
(2) If D is compound, say, $D = D_1 - D_2$, then $s \models D_1$ or $s \models D_2$.

In this way, we do not have to change the definition of $RF(D)$ for the definition of satisfaction, since what matters here is the set of representing facts of each constituent, which is atomic. Whether we need any change for comparing the semantic contents of diagrams, that is, whether two diagrams have equivalent representing facts or not, will be discussed in the next section.

Now, the definition for the consequence relation among diagrams is the same as in Venn-I:

[8]The following is the definition of the set of representing facts of diagram D, $RF(D)$, in Venn-I:

$$\begin{aligned} RF(D) &= \{\langle\langle \text{Shading}, x; 1\rangle\rangle \mid x \in RShading(D)\} \cup \{\langle\langle \otimes^n, y; 1\rangle\rangle \mid y \in R\otimes^n(D)\}, \\ \text{where } Seq(D) &= \langle BR(D),\ RShading(D),\ R\otimes^n(D)\rangle. \end{aligned}$$

For details, refer to §3.3.2.

Wfd D follows from set of *wfds* Δ (i.e., $\Delta \models D$) if and only if every set assignment that satisfies every member of Δ also satisfies D (i.e., $\forall_s(\forall_{D' \in \Delta} s \models D' \longrightarrow s \models D)$).

4.4 Equivalent diagrams and equivalent representing facts

In this section, we want to discuss syntactic equivalence and semantic equivalence among diagrams of Venn-II, as we did for Venn-I in §3.4.

4.4.1 Equivalent diagrams

First of all, the following definition of a sequence of a diagram will not function as it does in Venn-I:

$Seq(D) = \langle BR(D), RShading(D), R\otimes^n(D)\rangle$, where $BR(D)$ is the set of the basic regions of D, $RShading(D)$ is the set of the shaded regions of D, $R\otimes^n(D)$ is the set of the smallest regions[9] with an x-sequence of D.

The main reason why a sequence of a diagram is defined is to identify syntactically equivalent diagrams. However, if we kept the definition of equivalent diagrams from Venn-I without restricting the definition of a diagram sequence, the following (syntactically and semantically) different diagrams would turn out to be equivalent diagrams:

Example 47 *Let* $\langle A, B\rangle \notin cp$ *and* $\langle A, C\rangle$, $\langle B, D\rangle \in cp$.

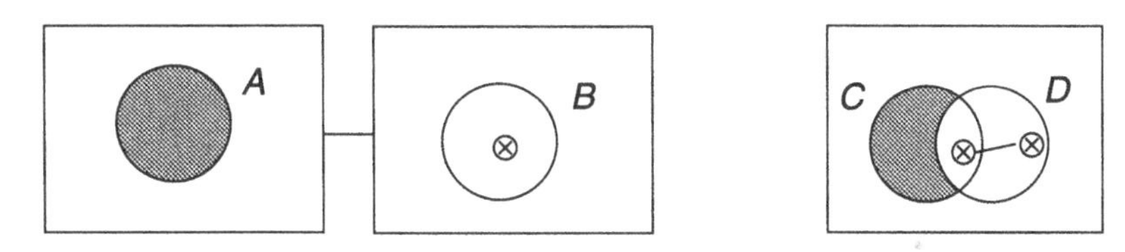

In order to avoid this result, we define the following syntactic relations among diagrams inductively, based upon the definitions of Venn-I.

Let D and D' be *wfds*.

$D \subseteq_{cp} D'$ (D is a *subdiagram* of D') iff either both D and D' are atomic and defined as in Venn-I,[10] or $D = D_1 - D_2$, $D' = D'_1 - D'_2$, $D_1 \subseteq_{cp} D'_1$, and $D_2 \subseteq_{cp} D'_2$.

[9]Refer to the earlier discussion about the ambiguity involving x-sequences.

[10]$D_1 \subseteq_{cp} D_2$ iff $BR(D_1) \subseteq_{cp} BR(D_2)$, $RShading(D_1) \subseteq_{cp} RShading(D_2)$, $R\otimes^n(D_1) \subseteq_{\equiv_{cp}} R\otimes^n(D_2)$, where $Seq(D_1) = \langle BR(D_1), RShading(D_1), R\otimes^n(D_1)\rangle$ and $Seq(D_2) = \langle BR(D_2), RShading(D_2), R\otimes^n(D_2)\rangle$.

$D \equiv_{cp} D'$ (D is *equivalent* to D') iff either both D and D' are atomic, $D \subseteq_{cp} D'$, and $D' \subseteq_{cp} D$, or $D = D_1\text{–}D_2$, $D' = D'_1\text{–}D'_2$, $D_1 \equiv_{cp} D'_1$, and $D_2 \equiv_{cp} D'_2$.

$D \subset_{cp} D'$ (D is a *proper subdiagram* of D') iff $D \subseteq_{cp} D'$ and $D \not\equiv_{cp} D'$.

(N.B.: If two diagrams D and D' have a different number of constituents, there is no inclusion or equivalence relation between D and D'.)

Example 48 *Let* $\langle A_1, A_4\rangle$, $\langle A_2, A_3\rangle \in cp$. *By definition,* $D_1 \equiv_{cp} D_4$ *and* $D_2 \equiv_{cp} D_3$. *However,* $D_1\text{–}D_2 \not\equiv_{cp} D_3\text{–}D_4$, *since* $D_1 \not\equiv_{cp} D_3$ *and* $D_2 \not\equiv_{cp} D_4$.

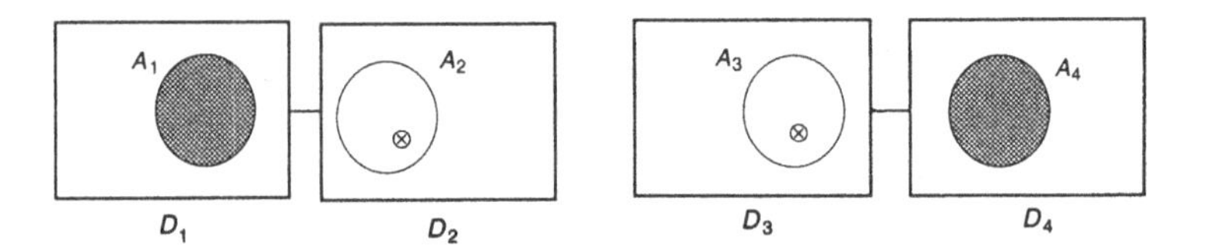

4.4.2 *Equivalent representing facts*

We can apply the same idea we used in the previous subsection to the comparison of representing facts.

Let D and D' be *wfd*s.

$RF(D) \subseteq_{cp} RF(D')$ iff either both D and D' are atomic and $\forall_{x \in RF(D)} \exists_{y \in RF(D')}(x \equiv_{cp} y)$, or $D = D_1\text{–}D_2$, $D' = D'_1\text{–}D'_2$, $RF(D_1) \subseteq_{cp} RF(D'_1)$, and $RF(D_2) \subseteq_{cp} RF(D'_2)$.

$RF(D) \equiv_{cp} RF(D')$ iff either both D and D' are atomic, $RF(D) \subseteq_{cp} RF(D')$, and $RF(D') \subseteq_{cp} RF(D)$, or $D = D_1\text{–}D_2$, $D' = D'_1\text{–}D'_2$, $RF(D_1) \equiv_{cp} RF(D'_1)$, and $RF(D_2) \equiv_{cp} RF(D'_2)$.

$RF(D) \subset_{cp} RF(D')$ iff $RF(D) \subseteq_{cp} RF(D')$ and $RF(D) \not\equiv_{cp} RF(D')$.

(N.B.: If two diagrams D and D' are different in the number of their constituents, there is no inclusion or equivalence relation between $RF(D)$ and $RF(D')$.)

In Venn-I, we have seen a difference between equivalent diagrams and equivalent representing facts. Through the following example, we know that neither equivalence among diagrams nor equivalence among representing facts is the same as logical equivalence.

Example 49 *Let* $\langle A_1, A_3\rangle$, $\langle A_2, A_4\rangle \in cp$. *Since* $D_1\text{–}D_2$ *and* D_3 *are different in the number of their constituents, there is no equivalence or inclusion* *relation between these diagrams or the representing facts of these diagrams. However,* $\{D_1–D_2\} \models D_3$ *and* $\{D_3\} \models D_1–D_2$.

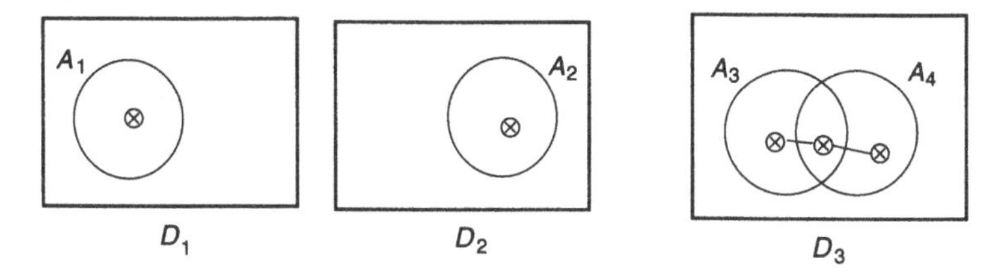

4.5 Transformation

There will be a change in the transformation rules in order to manipulate the new meaningful units, that is, compound diagrams.

4.5.1 Rules of transformation

Since we extended the set of well-formed diagrams, it is expected that we will have more rules in this extended system. We have six transformation rules in Venn-I. We keep four of them exactly the same as in Venn-I, modify one rule, the rule of conflicting information, and extend one rule, the unification rule. We add four more rules of transformation: the rule of splitting x's, the rule of the excluded middle, the rule of connecting a diagram, and the rule of construction. In the following, if the specification of a rule is exactly the same as in Venn-I, I will not repeat it.[11]

Rule 1: The rule of erasure of a diagrammatic object
Rule 2: The rule of erasure of part of an x-sequence
Rule 3: The rule of spreading x's

If *wfd* D has an x-sequence in a constituent, say, D_i, then we may copy D with ⊗ drawn in some other region of D_i and connected to the existing x-sequence.

Rule 4: The rule of introduction of a basic region
Rule 5: The rule of conflicting information

If *wfd* D has a region with both shading and an x-sequence in a constituent, say D_i, then we may copy D replacing D_i with any diagram we want.

[11] Refer to §3.5.1.

Rule 6: The rule of unification of diagrams

We may unify two diagrams D_1 and D_2 into one diagram, call it D, if each rectangle of D_1 stands in the counterpart relation to every rectangle of D_2.

We say that D is the *unification* of D' and D'' if the following conditions are satisfied:

(1) If D' and D'' are atomic *wfds*, then apply the rule of unification of diagrams in Venn-I.

(2) If $D' = D'_1 – D'_2$, then $D = D_1 – D_2$, where D_1 is the unification of D'_1 and D'' and D_2 is the unification of D'_2 and D''.

Example 50 *Let us unify the following diagrams:*

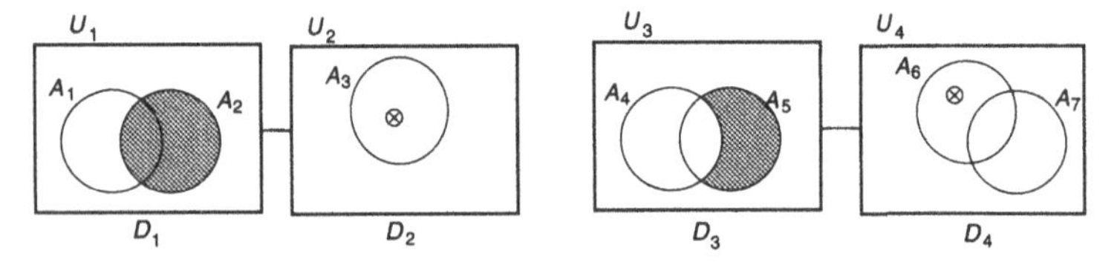

Let $\langle U_1, U_2\rangle$, $\langle U_2, U_3\rangle$, $\langle U_3, U_4\rangle$, $\langle A_1, A_4\rangle$, $\langle A_2, A_5\rangle$, $\langle A_2, A_7\rangle$, $\langle A_3, A_6\rangle \in cp$, *and* $\langle A_1, A_3\rangle$, $\langle A_2, A_3\rangle$, $\langle A_1, A_6\rangle$, $\langle A_3, A_7\rangle \notin cp$. *By the rule of unification of diagrams in Venn-I and the second clause of the rule of unification of diagrams in Venn-II, we get the following unified diagram:*

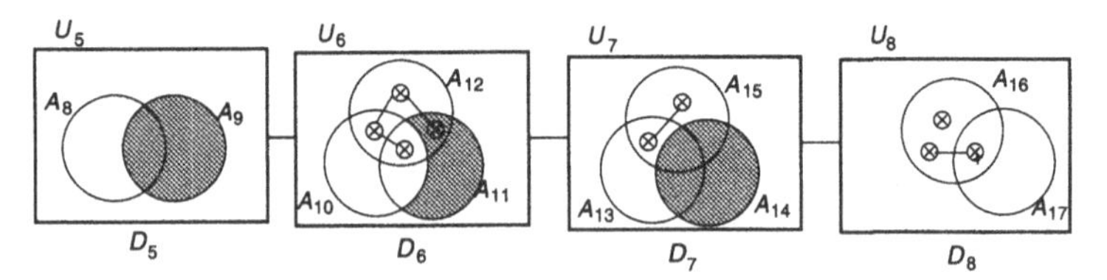

D_5 is a unification of D_1 and D_3, D_6 is that of D_2 and D_3, D_7 is that of D_1 and D_4, and D_8 is that of D_2 and D_4. Accordingly, $\langle U_5, U_6\rangle$, $\langle U_6, U_7\rangle$, $\langle U_7, U_8\rangle$, $\langle A_1, A_8\rangle$, $\langle A_2, A_9\rangle$, $\langle A_3, A_{12}\rangle$, $\langle A_4, A_{10}\rangle$, $\langle A_5, A_{11}\rangle$, $\langle A_6, A_{15}\rangle$, $\langle A_1, A_{13}\rangle$, $\langle A_2, A_{14}\rangle$, $\langle A_7, A_{17}\rangle$, $\langle A_3, A_{16}\rangle \in cp$.

The following are the new rules for this extended system.

Rule 7: The rule of splitting X's

Suppose D has constituent D_i such that $\langle\langle \otimes^n, A; 1\rangle\rangle \in RF(D_i)$, where $A = A_1 + \cdots + A_n$. Then, we may replace D_i with a connection of diagrams $D_i^1 – \cdots – D_i^n$ such that $D_i^1, \ldots, D_i^n$ are copies of D_i except that for every $D_i^j (i \leq j \leq n)$, $\langle\langle \otimes^n, A^*; 1 \rangle\rangle \notin RF(D_i^j)$, where $\langle A, A^* \rangle \in \overline{cp}$, and $\langle\langle \otimes^1, A'_1; 1 \rangle\rangle \in RF(D_i^1)$, ..., $\langle\langle \otimes^1, A'_n; 1 \rangle\rangle \in RF(D_i^n)$, where $\langle A_1, A'_1 \rangle$, ..., $\langle A_n, A'_n \rangle \in \overline{cp}$.

Example 51 *This rule allows us to transform the diagram on the left to the diagram on the right.*

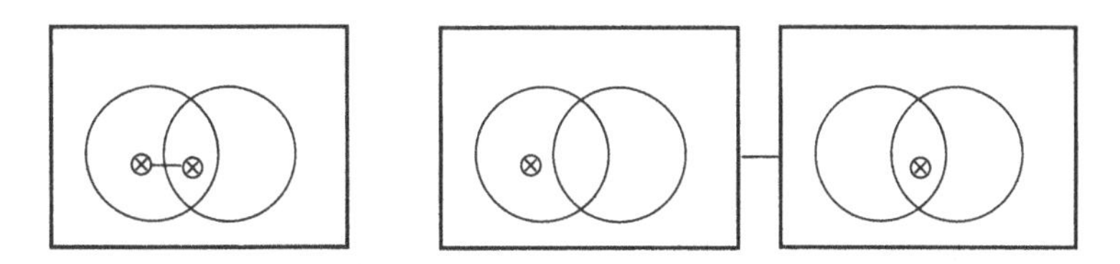

Rule 8: The rule of the excluded middle

Suppose D has constituent D_i such that for minimal region A of D_i, $\langle\langle \text{Shading}, A; 1 \rangle\rangle \notin RF(D_i)$ and $\langle\langle \otimes^1, A; 1 \rangle\rangle \notin RF(D_i)$. Then, we may replace D_i with D_i^1–D_i^2 such that D_i^1 and D_i^2 are copies of D_i except that $\langle\langle \text{Shading}, A_1; 1 \rangle\rangle \in RF(D_i^1)$ and $\langle\langle \otimes^1, A_2; 1 \rangle\rangle \in RF(D_i^2)$, where $\langle A, A_1 \rangle$, $\langle A, A_2 \rangle \in \overline{cp}$.

Example 52 *This rule allows us to transform the diagram on the left to the diagram on the right.*

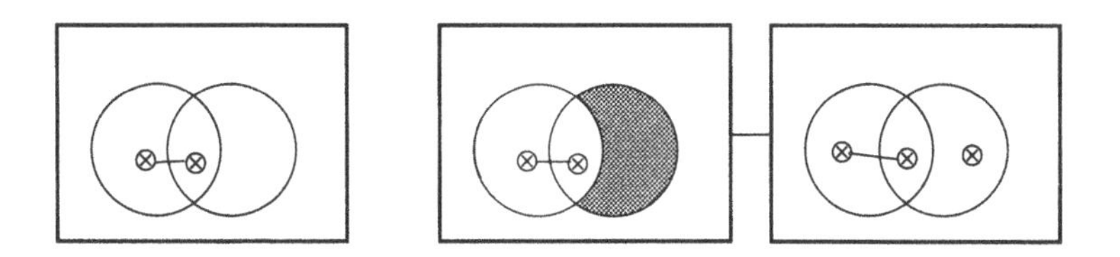

Rule 9: The rule of connecting a diagram

For a given diagram D, we may connect any diagram D' to D. That is, we may transform D into D'–D or D–D' for any D'.

Rule 10: The rule of construction

Given *wfd* D_1–$\cdots$–D_n, we may transform it into D if each of $D_1, \ldots, D_n$ may be transformed into D by some of the first nine transformation rules.

Example 53 *This rule allows us to transform the following first diagram into the second one:*

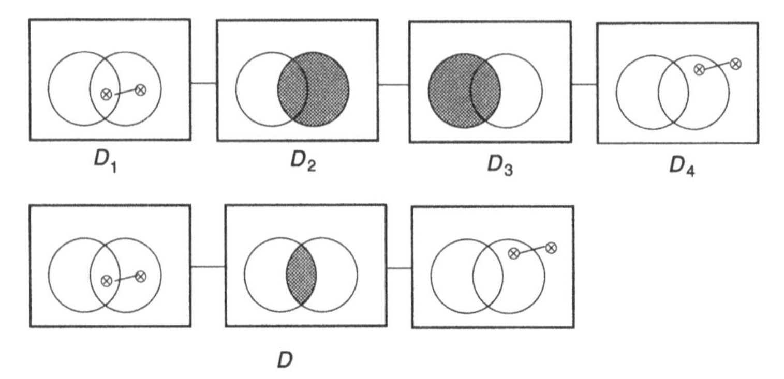

We may transform diagram D_1 into D by applying the rule of connecting a diagram twice. Diagram D_2 may be transformed into the following diagram by the rule of erasure of a diagrammatic object:

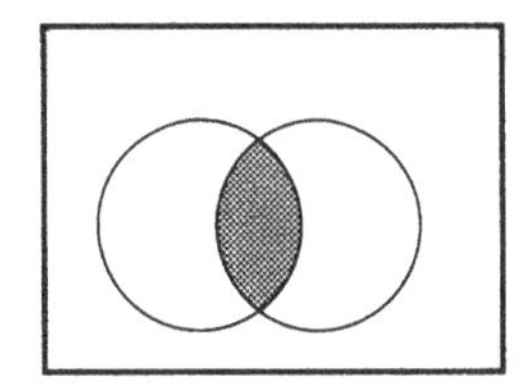

Again, we apply the rule of connecting a diagram twice to get diagram D. Similarly, we get the diagram D from both D_3 and D_4.

The following example shows us that this rule allows us to change the order of diagrams being connected.

Example 54 *We will see how the resulting diagram D of the previous example (i.e., example 53), is transformed into the following diagram:*

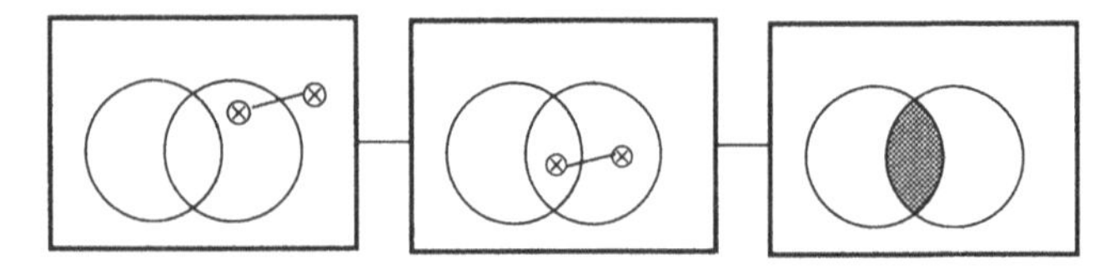

The first atomic diagram of D may be transformed into this diagram when we connect two atomic diagrams by the rule of connecting a diagram, one to its left side and the other to its right side. The same rule allows us to get this diagram from the other two atomic diagrams of D.

4.5.2 Formal obtainability

We define what it means for one diagram to be obtainable from other diagrams in the same way as in Venn-I. That is,

Definition *Let Δ be an empty set or a set of one wfd or a set of two wfds. Then, $\Delta \rightsquigarrow D$ iff there is a rule of transformation such that it allows us to transform the diagrams of Δ to D.*

Definition *Let Δ be a set of wfds and D be a wfd. Wfd D is obtainable from a set Δ of wfds ($\Delta \vdash D$) iff there is a sequence of wfds $\langle D_1, \ldots, D_n \rangle$ such that $D_n \equiv_{cp} D$ and for each k (where $1 \leq k \leq n$) either*

(a) there is some D' such that $D' \in \Delta$ and $D' \equiv_{cp} D_k$, or
(b) there is a set of diagrams Δ' such that for every diagram D_i, of Δ' $i < k$ and $\Delta' \rightsquigarrow D_k$.

4.6 Soundness

Let us prove the soundness of this extended system. First, we will prove that each transformation rule is valid.

4.6.1 Validity of transformation rules

In this subsection, I want to prove that each transformation rule is valid. For the validity of the first six rules, we will use the fact that the six transformation rules of Venn-I are valid.

Rule 1: The rule of erasure of a diagrammatic object

Let $\{D'\} \rightsquigarrow D$, where $D' = D'_1 - \cdots - D'_n$, $D = D_1 - \cdots - D_n$, and $\exists_{i(1 \leq i \leq n)}(\{D'_i\} \rightsquigarrow D_i)$ (by erasing a closed curve, an x-sequence or a shading).

Proof We know from Venn-I that $\{D'_i\} \models D_i$. Therefore, $\{D'\} \models D$. □

Rules 2–5: The proofs of these are similar to rule 1.
Rule 6: The rule of unification of diagrams

Let $\{D_1, D_2\} \rightsquigarrow D$ by rule 6.

Proof This follows by induction on the length of D_1.

(Basis case) Let D_1 be atomic. Suppose D_2 has k constituents, say, $D_2 = D_2^1 - \cdots - D_2^k$. That is,

$$\{D_1, D_2^1 - \cdots - D_2^k\} \rightsquigarrow D^1 - \cdots - D^k,$$

where $\{D_1, D_2^1\} \rightsquigarrow D^1, \ldots, \{D_1, D_2^k\} \rightsquigarrow D^k$.

We know from Venn-I that $\{D_1, D_2^1\} \models D^1, \ldots, \{D_1, D_2^k\} \models D^k$. That is,

$$\forall_s((s \models D_1 \ \wedge \ s \models D_2^1) \longrightarrow s \models D^1),$$

$$\vdots$$

$$\forall_s((s \models D_1 \ \wedge \ s \models D_2^k) \longrightarrow s \models D^k).$$

Accordingly,

$$\forall_s((s \models D_1 \ \wedge \ (s \models D_2^1 \ \vee \ \cdots \ \vee s \models D_2^k)) \longrightarrow (s \models D^1 \ \vee \ \cdots \ \vee s \models D^k)).$$

Therefore, $\{D_1, D_2^1 - \cdots - D_2^k\} \models D^1 - \cdots - D^k$.

(Inductive step) Suppose that D_1 has n constituents. That is, $D_1 = D_1^1 - \cdots - D_1^n$. Let $D^* = D_1^1 - \cdots - D_1^{n-1}$. That is, $\{D^* - D_1^n, D_2\} \rightsquigarrow D^1 - D^2$ (by the rule of unification), where $\{D^*, D_2\} \rightsquigarrow D^1$ and $\{D_1^n, D_2\} \rightsquigarrow D^2$. By the inductive hypothesis, $\{D^*, D_2\} \models D^1$ and $\{D_1^n, D_2\} \models D^2$. Therefore, $\{D^* - D_1^n, D_2\} \models D^1 - D^2$. □

Rule 7: The rule of splitting X's

Let $\{D'\} \rightsquigarrow D$, where D_i is a constituent of D', $D_i^1, \ldots, D_i^n$ are constituents of D, and $\{D_i\} \rightsquigarrow D_i^1 - \cdots - D_i^n$ by rule 7.

Proof Suppose $\langle\!\langle \otimes^n, A; 1\rangle\!\rangle \in RF(D_i)$, where $A = A_1 + \cdots + A_n$ and $\langle\!\langle \otimes^1, A_1'; 1\rangle\!\rangle \in RF(D_i^1)$, ..., $\langle\!\langle \otimes^1, A_n'; 1\rangle\!\rangle \in RF(D_i^n)$, where $\langle A_1, A_1'\rangle$, ..., $\langle A_n, A_n'\rangle \in \overline{cp}$. But, since $\forall_s(\overline{s}(A) = \overline{s}(A_1) \cup \cdots \cup \overline{s}(A_n))$,

$$\forall_s(s \models \langle\!\langle \otimes^n, A; 1\rangle\!\rangle \longrightarrow (s \models \langle\!\langle \otimes^1, A_1'; 1\rangle\!\rangle \ \vee \ \cdots \ \vee s \models \langle\!\langle \otimes^1, A_n'; 1\rangle\!\rangle)).$$

Accordingly, $\{D_i\} \models D_i^1 - \cdots - D_i^n$. Therefore, $\{D'\} \models D$. □

Rule 8: The rule of the excluded middle

Let $\{D'\} \rightsquigarrow D$, where D_i is a constituent of D', D_i^1 and D_i^2 are constituents of D, and $\{D_i\} \rightsquigarrow D_i^1 - D_i^2$ by rule 8.

Proof Suppose $\langle\!\langle \text{Shading}, A; 1\rangle\!\rangle \notin RF(D_i)$, $\langle\!\langle \otimes^n, A; 1\rangle\!\rangle \notin RF(D_i)$, $\langle\!\langle \text{Shading}, A_1; 1\rangle\!\rangle \in RF(D_i^1)$, and $\langle\!\langle \otimes^1, A_2; 1\rangle\!\rangle \in RF(D_i^2)$, where $\langle A, A_1\rangle$, $\langle A, A_2\rangle \in \overline{cp}$. But, for every pair of regions x, y such that $\langle x, y\rangle$

$\in \overline{cp}$, $\forall_s(\bar{s}(x) = \emptyset \vee \bar{s}(y) \neq \emptyset)$. Accordingly, $\forall_s(s \models \langle\langle \text{Shading}, x; 1\rangle\rangle \vee s \models \langle\langle \otimes^n, y; 1\rangle\rangle)$. Accordingly, $\{D_i\} \models D_i^1\text{–}D_i^2$. Therefore, $\{D'\} \models D$. □

Rule 9: The rule of connecting a diagram

Let $\{D'\} \rightsquigarrow D'\text{–}D$ or $\{D'\} \rightsquigarrow D\text{–}D'$ for any diagram D.

Proof For any diagram D, $\{D'\} \models D'\text{–}D$ and $\{D'\} \models D\text{–}D'$. □

Rule 10: The rule of construction

Suppose that $\{D_1\} \vdash D, \ldots, \{D_n\} \vdash D$ by some of the first nine rules.

Proof Since we proved that these nine transformation rules are valid, we can show easily that $\{D_1\} \models D, \ldots, \{D_n\} \models D$. Hence, obviously, $\{D_1\text{–}\cdots\text{–}D_n\} \models D$. □

4.6.2 Soundness theorem

Soundness Theorem *If* $\Delta \vdash D$, *then* $\Delta \models D$.

Proof This proof is the same as the proof in §3.6.2.

4.7 Completeness

4.7.1 Maximal diagram

In the case of Venn-I, we developed a notion of maximal diagram to prove the completeness of the system. We want to introduce a very similar idea to the Venn-II system as well. There are two reasons why we need a slight modification of the definition of maximal diagram in Venn-I for the definition of maximal diagram in Venn-II. One is that the relations of basic regions of diagrams are not easily compared, since a diagram might be compound and we did not define the set of basic regions of a compound diagram. Second, since we introduced a connected diagram, for a given diagram we can always connect another diagram with this given diagram so that the new one is a consequence of the original one. That is, any diagram with D as its constituent is a consequence of D. It is impossible to draw a diagram that includes all these diagrams. Hence, we apply the concept of a maximal diagram of Venn-I only to the constituents of a diagram of Venn-II, which are atomic. Given D, we construct its maximal diagram, say, D^*, in such a way that each constituent of D^* is a maximal (in the sense of Venn-I) of its corresponding constituent of D. The following inductive definition reflects this idea:

Definition *D^* is a maximal of D iff*

(1) *if D is atomic, then*

$$\forall_{atomicD'}((\{D\} \models D' \wedge BR(D') \subseteq_{cp} BR(D)) \longleftrightarrow D' \subseteq_{cp} D^*);$$

(2) *if D is compound, say, $D_1 - D_2$, then $D^* = D_1^* – D_2^*$, where D_1^* is a maximal of D_1 and D_2^* is a maximal of D_2.*

There are several important features of maximal diagrams. First, if D is atomic, its maximal diagram is also atomic. Second, diagram D and its maximal diagram have the same number of constituents. Third, if D^* is maximal, then every constituent of D^* is also maximal.

We can prove the existence and the uniqueness of a maximal diagram as we did in Venn-I.

Theorem 18 (Maximal Representation Theorem) *For every diagram D, there is a maximal diagram D^* of D such that $\{D\} \vdash D^*$.*

Proof This follows by induction on the length of D. □

Corollary 18 *For every nonempty set Δ of diagrams,*[12] *there is a diagram D^* such that $\Delta \vdash D^*$ and D^* is a maximal of D, where D is the unification of the diagrams in Δ.*

Definition *We call the diagram D^* in Corollary 18 a maximal diagram of Δ.*

Theorem 19 (Uniqueness of a Maximal Diagram) *Let D, D_1, and D_2 be wfds and cp be a given counterpart relation. If D_1 and D_2 are maximal diagrams of D, then $D_1 \equiv_{cp} D_2$.*

Theorem 20 *Let D^* be a maximal of D. Then, D and D^* are logical consequences of each other. (I.e., $\{D^*\} \models D$ and $\{D\} \models D^*$.)*

[12] As in Venn-I, I will limit myself to a finite set of diagrams.

4.7.2 Elaborated maximal diagram

In this subsection, we are going to introduce two new notions: elaborated diagram and elaborated maximal diagram. In Venn-II, there are two kinds of syntactic devices to represent disjunctive information: connecting ⊗'s by a line and connecting diagrams by a line. We have the former one in Venn-I as well. However, as seen in the first part of this chapter, this syntactic device is not enough to express every kind of disjunctive information using categorical sentences. On the other hand, the information that connected ⊗'s convey may always be expressible in terms of connected diagrams, since we have a rule of splitting x's, which is proven to be valid. Here we are interested in converting one kind of syntactic device to the other kind so that each constituent of a diagram has only conjunctive information. Let me illustrate this process in the following example:

Example 55 *In this following diagram, there is one x-sequence with two ⊗'s.*

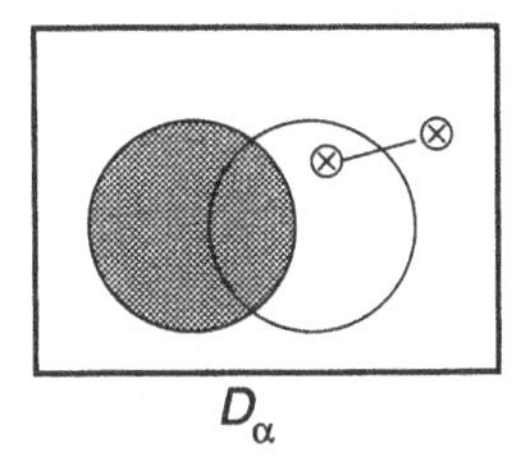

We apply the rule of splitting x's and we obtain the following diagram:

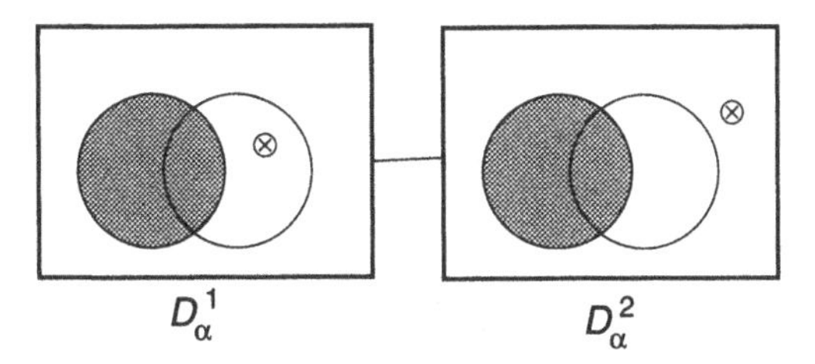

Each constituent of this new diagram D_α^1–D_α^2 has only conjunctive information. In that sense, this diagram seems to correspond to a disjunctive normal form in symbolic logic. In the case of symbolic logic, if α' is a disjunctive normal form of α, then α and α' are logically equivalent. Interestingly, the

following is true as well:

$$\{D_\alpha\} \models D^1_\alpha\text{–}D^2_\alpha, \tag{1}$$

since $\{D_\alpha\} \vdash D^1_\alpha\text{–}D^2_\alpha$ *by the rule of splitting x's.*

$$\{D^1_\alpha\text{–}D^2_\alpha\} \models D_\alpha, \tag{2}$$

since $\{D^1_\alpha\text{–}D^2_\alpha\} \vdash D_\alpha$ *by the rule of spreading x's and the rule of construction.*

In chapter 2, we noticed that Venn improved Euler's diagrams to represent our imperfect knowledge about the domain. That uncertainty is represented by an empty region. By an empty region, I mean a region that has neither shading nor an x-sequence. The rule of the excluded middle may be applied to an empty region. Hence, we may apply the rule of the excluded middle until we do not have any empty region. Interestingly, the original diagram and the resulting diagram are logically equivalent to each other. Let us continue this process with the previous diagram, $D^1_\alpha\text{–}D^2_\alpha$. That is, we are interested in transforming $D^1_\alpha\text{–}D^2_\alpha$ to a diagram, say, D', satisfying the following:

(1) Each minimal region of D' has either a shading or an ⊗.
(2) $D^1_\alpha\text{–}D^2_\alpha$ and D' are logically equivalent to each other.

In order to satisfy the first condition, let us apply the rule of the excluded middle to each empty minimal region of $D^1_\alpha\text{–}D^2_\alpha$ to get the following:

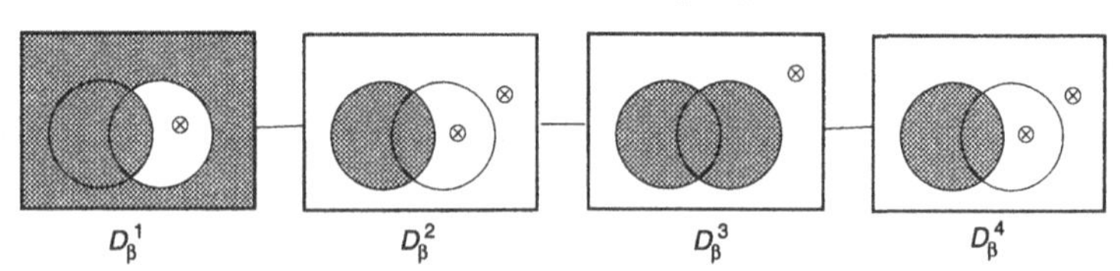

Also, the second condition is satisfied, since the following is true as well: (Let $D' = D^1_\beta\text{–}D^2_\beta\text{–}D^3_\beta\text{–}D^4_\beta$.)

$$\{D^1_\alpha\text{–}D^2_\alpha\} \models D', \tag{1}$$

since $\{D^1_\alpha\text{–}D^2_\alpha\} \vdash D'$ (by the rule of the excluded middle).

$$\{D'\} \models D^1_\alpha\text{–}D^2_\alpha, \tag{2}$$

since $\{D'\} \vdash D^1_\alpha\text{–}D^2_\alpha$ (by the rule of erasure of a diagrammatic object and by the rule of construction).

We call D' the *elaborated diagram* of D_α, or simply the E-diagram. An E-diagram has the following features: First, there is no x-sequence with more than one ⊗. That is, each constituent has only conjunctive information. Second, there is no minimal region that is empty, that is, not shaded and does not have an x-sequence. Third, a diagram and its E-diagram are logically equivalent to each other.

Now we want to obtain a maximal diagram of this E-diagram, and call it the elaborated maximal diagram (simply, E-maximal diagram) of D_α. That is, we make a maximal diagram for each atomic diagram of D^1_β–D^2_β–D^3_β–D^4_β. (I will leave it to the reader.)

For a given diagram, say, D, let us generalize the process to get its E-maximal diagram, say, D^{E*}:

(Step 1) For each x-sequence with two or more ⊗'s (i.e., $\otimes^n$, where $2 \leq n$), apply the rule of splitting x's. That is,

$$\{D\} \vdash D^1 - \cdots - D^m. \tag{1}$$

(Step 2) For each D^i (where $1 \leq i \leq m$), apply the rule of the excluded middle until every minimal region has either shading or an x-sequence. That is,

$$\begin{gathered} \{D^1\} \vdash D^1_\alpha - \cdots - D^j_\alpha, \\ \vdots \\ \{D^m\} \vdash D^k_\alpha - \cdots - D^r_\alpha. \end{gathered} \tag{2}$$

Apply the rule of connecting a diagram and the rule of construction to get the following:

$$\{D^1 - \cdots - D^m\} \vdash D^1_\alpha - \cdots - D^r_\alpha. \tag{3}$$

Diagram $D^1_\alpha - \cdots - D^r_\alpha$ is the E-diagram of D.

(Step 3) We get a maximal diagram for each D^i_α and call it D^i_*. That is,

$$\{D^1_\alpha - \cdots - D^r_\alpha\} \vdash D^1_* - \cdots - D^r_* \tag{4}$$

By equations (1), (3), and (4),

$$\{D\} \vdash D^1_* - \cdots - D^r_*. \tag{5}$$

Let $D^{E*} = D^1_* - \cdots - D^r_*$ and call it the E-maximal of D.

We want to prove that D and its E-maximal diagram are logically equivalent to each other. We showed informally that D and its E-diagram are logically equivalent to each other and we know that a diagram and its maximal diagram are logically equivalent. Therefore, it should be the case that D and its E-maximal diagram are logically equivalent. For a given diagram D, its E-maximal is obtained by the steps just mentioned. Therefore, by soundness, we know that the E-maximal of D is a consequence of D. Let us prove the following theorem:

Theorem 21 *Let* D^{E*} *be the E-maximal of diagram D. Then,* $\{D^{E*}\} \models D$.

Proof Check each step we followed for constructing an E-maximal diagram of D.

(Step 1) Notice that we may transform diagram D^1–$\cdots$–D^m to get D by applying the rule of spreading x's. That is,

$$\{D^1 - \cdots - D^m\} \vdash D, \tag{6}$$

$$\{D^1 - \cdots - D^m\} \models D \text{ (by soundness).} \tag{7}$$

(Step 2) Notice that we may transform diagram D^1_α–$\cdots$–D^r_α to obtain D^1–$\cdots$–D^m by applying the rule of erasing diagrammatic objects (i.e., shadings and x-sequences) and by the rule of construction. That is,

$$\{D^1_\alpha - \cdots - D^r_\alpha\} \vdash D^1 - \cdots - D^m, \tag{8}$$

$$\{D^1_\alpha - \cdots - D^r_\alpha\} \models D^1 - \cdots - D^m \text{ (by soundness).} \tag{9}$$

(Step 3) Theorem 13 tells us that $\{D^1_*\} \models D^1_\alpha, \ldots, \{D^r_*\} \models D^r_\alpha$. Therefore,

$$\{D^{E*}\} \models D^1_\alpha - \cdots - D^r_\alpha, \text{ where } D^{E*} = D^1_* - \cdots - D^r_*. \tag{10}$$

By equations (7), (9), and (10),

$$\{D^{E*}\} \models D. \tag{11}$$

□

4.7.3 Completeness theorem

The main idea of the completeness proof of Venn-I is as follows: Suppose diagram D is a consequence of set Δ of diagrams. We get the maximal diagram of Δ. By the definition of maximal diagram, we know that this maximal diagram includes all and only diagrams that follow from the diagrams of Δ. Therefore, we may erase diagrammatic objects (by the rule of erasure of diagrammatic objects) or introduce basic regions (by the rule of introduction of a basic region) until we get a diagram that is equivalent to D.

This strategy does not work for the completeness proof of Venn-II. First of all, as we discussed before, the definition of maximal diagram has to be modified in this extended system. Therefore, we cannot get the maximal diagram such that it includes all and only diagrams that are consequences of given diagrams. Instead, we use the notion of E-maximal diagram defined in the previous subsection.

Before going through the proof of completeness, let me illustrate the basic idea of the most complicated part of the proof by the following example:

Example 56 *In the following,* $\{D'\} \models D_1\text{–}D_2$.

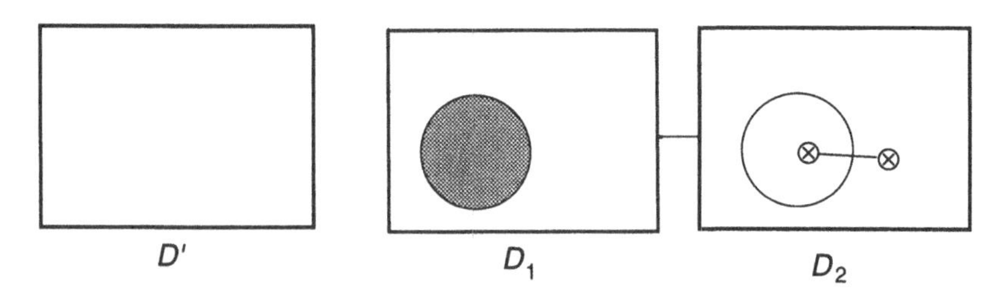

The question is whether $\{D'\} \vdash D_1\text{–}D_2$ or not. The maximal diagram of D' would not help us much, since the maximal diagram of D' is equivalent to D'. That is, it also consists of only a rectangle. There are two things we should notice; One is that $\{D'\} \not\models D_1$, and $\{D'\} \not\models D_2$ but $\{D'\} \models D_1\text{–}D_2$. The other is that $BR(D') \not\equiv_{cp} BR(D_1)$ and $BR(D') \not\equiv_{cp} BR(D_2)$. My proof suggests the following:

1. Apply the rule of introduction of a closed curve to D':

$$\{D'\} \vdash D_\alpha, \text{ where } BR(D_\alpha) \equiv_{cp} BR(D_1) \text{ and } BR(D_\alpha) \equiv_{cp} BR(D_2).$$

That is,

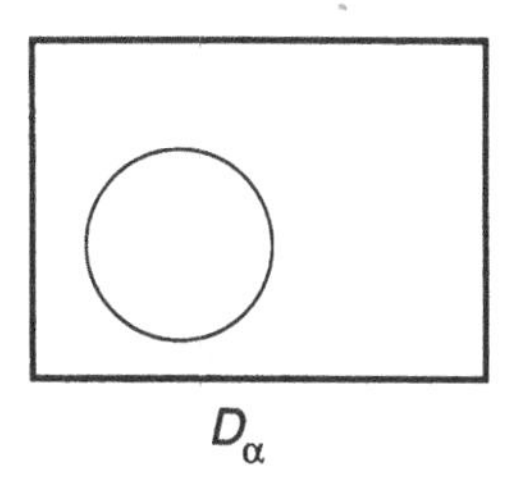

2. Obtain the E-maximal diagram D^{E*} from D_α:

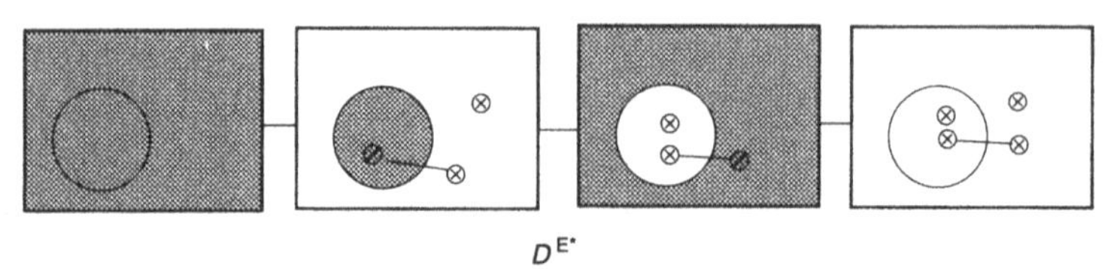

I claim that each atomic diagram of D^{E^*} includes (i.e., $\subseteq_{cp}$) at least one atomic diagram of D_1–D_2, which is true in this case. The first constituent of D^{E^*} includes D_1, the second one D_2, the third one D_2, and the fourth one D_2. Accordingly, we can obtain each constituent of D_1–D_2 from this E-maximal diagram by erasing diagrammatic objects. Therefore, it is the case that $\{D'\} \vdash D_1$–D_2.

We need to prove two more theorems for the completeness proof.

Theorem 22 *Suppose the following is true, where every diagram is atomic:*

(1) $\{D\} \models D_1 - \cdots - D_k$ $(2 \leq k)$.

(2) *For every* D', $\{D\} \not\models D'$, *where* D' *is the same as* $D_1 - \cdots - D_k$ *except that one atomic diagram,* $D_i (1 \leq i \leq k)$, *is removed from* $D_1 - \cdots - D_k$. *That is,* $D' = D_1 - \cdots - D_{i-1} - D_{i+1} - \cdots - D_k$ *for some* i $(1 \leq i \leq k)$.

(3) *For every* i $(\leq i \leq k)$ $\forall_{D_i}(BR(D_i) \subseteq_{cp} BR(D))$.

(Note: D does not have any region both with a shading and with an x-sequence, since if it did, clause 2 would not be true.)

Then (1) for every region A_1 *such that* $\langle\langle Shading, A_1; 1\rangle\rangle \in RF(D_i)$ *for some* i *(where* $1 \leq i \leq k$*),* $\langle\langle \otimes^n, A_2; 1\rangle\rangle \notin RF(D)$, *where* A_2 *is a region of* D *and* $\langle A_1, A_2\rangle \in \overline{cp}$.

(2) For every region A_1 *such that* $\langle\langle \otimes^n, A_1; 1\rangle\rangle \in RF(D_i)$ *for some* i *(where* $1 \leq i \leq k$*),* $\langle\langle Shading, A_2; 1\rangle\rangle \notin RF(D)$, *where* A_2 *is a region of* D *and* $\langle A_1, A_2\rangle \in \overline{cp}$.

Proof (1) Suppose that $\langle\langle$Shading, $A_1; 1\rangle\rangle \in RF(D_i)$. By assumption (3), we know that there is a region in D such that $\langle A_1, A_2\rangle \in \overline{cp}$. Suppose that $\langle\langle \otimes^n, A_2; 1\rangle\rangle \in RF(D)$. Let D' be the same as $D_1 - \cdots - D_k$ except that one atomic diagram, $D_i(1 \leq i \leq k)$, is removed from it. By assumption (2),

$$\exists_s(s \models D \wedge s \not\models D'). \tag{4}$$

Let s_1 be such that

$$s_1 \models D \wedge s_1 \not\models D'. \tag{5}$$

Since we assumed that $\langle\langle \otimes^n, A_2; 1\rangle\rangle \in RF(D)$, and $s_1 \models D$ (by equation (5)),

$$\overline{s_1}(A_1) \neq \emptyset. \tag{6}$$

(N.B.: $\langle A_1, A_2\rangle \in \overline{cp}$). By equation (6) and our assumption that $\langle\langle$Shading, $A_1; 1\rangle\rangle \in RF(D_i)$,

$$s_1 \not\models D_i. \tag{7}$$

Therefore, by equation (5) and (7),

$$s_1 \models D \ \wedge \ s_1 \not\models D_1 \ \wedge \ \cdots \ \wedge \ s_1 \not\models D_k. \tag{8}$$

Equation (8) contradicts our assumption (1), and thus our first claim is proved.

(2) Suppose that $\langle\langle \otimes^n, A_1; 1\rangle\rangle \in RF(D_i)$. By assumption (3), we know that there is a region in D such that $\langle A_1, A_2\rangle \in \overline{cp}$. Suppose that $\langle\langle$Shading, $A_2; 1\rangle\rangle \in RF(D)$. Let D' be the same as $D_1 - \cdots - D_k$ except that one atomic diagram, $D_i (1 \leq i \leq k)$, is removed from it. By assumption (2),

$$\exists_s (s \models D \wedge s \not\models D'). \tag{9}$$

Let s_1 be such that

$$s_1 \models D \wedge s_1 \not\models D'. \tag{10}$$

Since we assumed that $\langle\langle$Shading, $A_2; 1\rangle\rangle \in RF(D)$, and $s_1 \models D$ (by equation (10)), we know that

$$\overline{s_1}(A_2) = \emptyset. \tag{11}$$

Since $\langle A_1, A_2\rangle \in \overline{cp}$, by equation (11),

$$\overline{s_1}(A_1) = \emptyset. \tag{12}$$

By equation (12) and our assumption that $\langle\langle \otimes^n, A_1; 1\rangle\rangle \in RF(D_i)$,

$$s_1 \not\models D_i. \tag{13}$$

Therefore, by equations (10) and (13),

$$s_1 \models D \ \wedge \ s_1 \not\models D_1 \ \wedge \ \cdots \ \wedge \ s_1 \not\models D_k. \tag{14}$$

Equation (14) contradicts our assumption (1). □

Theorem 23 *If $\{D'\} \models D$, then for every constituent of D', say, D'_i, $\{D'_i\} \models D$.*

Proof It follows by induction of the length of D'.

(Basis case) Let D' be an atomic diagram. Since $D'_i = D'$, obviously $\{D'_i\} \models D$.

(Inductive step) Let D' have n atomic diagrams. Let $D' = D_1 – D_2$. That is, $\{D_1 – D_2\} \models D$. Accordingly, $\forall_s((s \models D_1 \ \vee s \models D_2) \rightarrow s \models D)$. That is,

$$\forall_s(s \models D_1 \rightarrow s \models D) \wedge \forall_s(s \models D_2 \rightarrow s \models D).$$

By the inductive hypothesis, D follows from every constituent of D_1 and from every constituent of D_2. Since every constituent of D' is a constituent either of D_1 or of D_2, D follows from every constituent of D'. □

Completeness: *If* $\Delta \models D$, *then* $\Delta \vdash D$.

Proof If $\Delta = \emptyset$, then $\Delta \vdash D^0$ by the rule of introduction of a basic region, where D^0 is a diagram that consists only of one rectangle. By the tautology lemma (in §3.7.2), $\{D^0\} \models D$. Let D' be a maximal of D^0. Accordingly, $\{D^0\} \vdash D'$. Hence, $\Delta \vdash D'$. Since $\{D'\} \models D^0$, $\{D'\} \models D$. If $\Delta \neq \emptyset$ and D' is a maximal of Δ, then $\Delta \vdash D'$. Since $\forall_{D_x \in \Delta}\{D'\} \models D_x$ and $\Delta \models D$ (by our assumption), $\{D'\} \models D$. In either case, we get the following:

$$\{D'\} \models D, \tag{1}$$

$$\Delta \vdash D'. \tag{2}$$

We want to show $\{D'\} \vdash D$ by induction on the length of D.

(Basis case) Let D be atomic. Suppose $D' = D'_1 – \cdots – D'_m$. By equation (1) and Theorem 23, for each constituent of D',

$$\{D'_i\} \models D \ (1 \leq i \leq m). \tag{3}$$

Since D' is maximal, each constituent is also maximal. By (3) and Theorem 16, we know that there is a diagram D^*_α such that

$$\{D'_i\} \vdash D^*_\alpha, \tag{4}$$

$$D \subseteq_{cp} D^*_\alpha. \tag{5}$$

By equation (5) and Theorem 9,

$$\{D^*_\alpha\} \vdash D. \tag{6}$$

By equations (4) and (6),

$$\{D_i'\} \vdash D \ (1 \leq i \leq m). \tag{7}$$

By applying the rule of construction, we get the following:

$$\{D'\} \vdash D. \tag{8}$$

(Inductive step) Inductive hypothesis: If D has k (where $k < n$) atomic diagrams as its constituents, then $\{D'\} \vdash D$.

Suppose that D has n constituents. That is, $D = D_1-\cdots-D_n$. Is there any constituent D_i of D such that $\{D'\} \models D_1-\cdots-D_{i-1}-D_{i+1}-\cdots-D_n$?

(Case 1) Yes. By the rule of connecting diagrams,

$$\{D_1-\cdots-D_{i-1}\} \vdash D_1-\cdots-D_{i-1}-D_i-\cdots-D_n,$$
$$\{D_{i+1}-\cdots-D_n\} \vdash D_1-\cdots-D_1^i-D_{i+1}-\cdots-D_n.$$

By the rule of construction,

$$\{D_1-\cdots-D_{i-1}-D_{i+1}-\cdots D_n\} \vdash D_1-\cdots-D_n.$$

By the inductive hypothesis, we know that

$$\{D'\} \vdash D_1-\cdots-D_{i-1}-D_{i+1}-\cdots D_n.$$

By transitivity, we get the following:

$$\{D'\} \vdash D.$$

(Case 2) No. We check for each constituent of D', say, D_i', whether or not there is an atomic diagram of D, say, D_j, such that $\{D_i'\} \models D_j$.

(1) Yes. Since D_i' is maximal, as seen in the basis case, $\{D_i'\} \vdash D_j$.
(2) No. But, since $\{D'\} \models D$, by Theorem 23 $\{D_i'\} \models D$. Choose the shortest diagram from D that follows from D_i'. That is,

$$\{D_i'\} \models D_\gamma^1-\cdots-D_\gamma^k, \tag{9}$$

where $D_\gamma^j (1 \leq j \leq k)$ is a constituent of D;

$$\forall_{D_\gamma^j(1\leq j\leq k)}(\{D_i'\} \not\models D_\gamma^1-\cdots-D_\gamma^{j-1}-D_\gamma^{j+1}-\cdots-D_\gamma^k). \tag{10}$$

Is it the case that $\forall_{D_\gamma^j(1\leq j\leq k)}(BR(D_\gamma^j) \subseteq_{cp} BR(D_i'))$?

(2-i) Yes. Let $D_i^\alpha = D_i'$.
(2-ii) No. Transform D_i' by the rule of introduction of a closed curve,

$$\{D_i'\} \vdash D_i^\alpha, \text{ where } \forall_{D_\gamma^j(1\le j\le k)}(BR(D_\gamma^j) \subseteq_{cp} BR(D_i^\alpha)). \tag{11}$$

Transform D_i^α, obtained from either (2-i) or (2-ii), to get its maximal, say, $D_i^{\alpha *}$. That is,

$$\{D_i^\alpha\} \vdash D_i^{\alpha *}, \tag{12}$$

where $D_i^{\alpha *}$ is the maximal of D_i^α.

Transform $D_i^{\alpha *}$ to get its E-maximal diagram. Call it $D_{\beta *}$. Let $D_{\beta *} = D_{\beta *}^1 - \cdots - D_{\beta *}^m$. That is,

$$\{D_i^{\alpha *}\} \vdash D_{\beta *}^1 - \cdots - D_{\beta *}^m. \tag{13}$$

By Theorem 21,

$$\{D_{\beta *}^1 - \cdots - D_{\beta *}^m\} \models D_i^{\alpha *}. \tag{14}$$

By equation (12) and Theorem 20,

$$\{D_i^{\alpha *}\} \models D_i^\alpha. \tag{15}$$

In (2-i), $D_i^\alpha = D_i'$. In (ii-2), we may obtain D_i' from D_i^α by erasing closed curves. In either case,

$$\{D_i^\alpha\} \models D_i'. \tag{16}$$

By equations (14), (15), (16), and (9),

$$\{D_{\beta *}^1 - \cdots - D_{\beta *}^m\} \models D_\gamma^1 - \cdots - D_\gamma^k. \tag{17}$$

Claim *For every atomic diagram of this E-maximal diagram, say, $D_{\beta *}^l$ ($1 \le l \le m$), there is an atomic diagram, say, D_γ^j ($1 \le j \le k$), of $D_\gamma^1 - \cdots - D_\gamma^k$ such that $\{D_{\beta *}^l\} \models D_\gamma^j$.*

Proof of Claim: Suppose that there is an atomic diagram of $D_{\beta *}$, say, $D_{\beta *}^l$, such that $\neg\exists_{D_\gamma^j}(\{D_{\beta *}^l\} \models D_\gamma^j)$. By equation (17), we know that there exist diagrams $D_\alpha^1, \ldots, D_\alpha^p$, which are constituents of $D_\gamma^1 - \cdots - D_\gamma^k$, satisfying the following: (N.B.: By our supposition, we know $2 \le l$.)

$$\{D_{\beta *}^l\} \models D_\alpha^1 - \cdots - D_\alpha^p, \tag{18}$$

where each D_α^j is a constituent of $D_\gamma^1 - \cdots - D_\gamma^k$.

$$\forall_{D_\alpha^j(1\le j\le p)}(\{D_{\beta *}^l\} \not\models D_\alpha^1 - \cdots - D_\alpha^{j-1} - D_\alpha^{j+1} - \cdots - D_\alpha^p). \tag{19}$$

Since $\forall_{D_\gamma^j(1\le j\le k)}(BR(D_\gamma^j) \subseteq_{cp} BR(D_i^\alpha))$ and $BR(D_i^\alpha) \equiv_{cp} BR(D_{\beta *}^l)$,

$$\forall_{D_\alpha^j(1\le j\le p)}(BR(D_\alpha^j) \subseteq_{cp} BR(D_{\beta *}^l)). \tag{20}$$

For each D_α^j, there are two cases (by our assumption):

(i) $\langle\!\langle \text{Shading}, A_1; 1\rangle\!\rangle \in RF(D^j_\alpha)$, but $\langle\!\langle \text{Shading}, A_2; 1\rangle\!\rangle \notin RF(D^l_{\beta*})$, where $\langle A_1, A_2\rangle \in \overline{cp}$.

Claim 1 of Theorem 22 says that $\langle\!\langle \otimes^n, A_2; 1\rangle\!\rangle \notin RF(D^l_{\beta*})$, where $\langle A_1, A_2\rangle \in \overline{cp}$. If A_2 is a minimal region, it is impossible for there to be a minimal region of $D^l_{\beta*}$ with no shading or with no x-sequence (by the construction of the E-maximal diagram). If A_2 is not a minimal region, since $\langle\!\langle \text{Shading}, A_2; 1\rangle\!\rangle \notin RF(D^l_{\beta*})$, there is a minimal region A' satisfying the following:

$$A' \text{ is part of } A_2. \tag{21}$$

$$\langle\!\langle \text{Shading}, A'; 1\rangle\!\rangle \notin RF(D^l_{\beta*}). \tag{22}$$

Since $D^l_{\beta*}$ is maximal and $D^l_{\beta*} \not\models \langle\!\langle \otimes^n, A_2; 1\rangle\!\rangle$, we know that no minimal region that is part of A_2 has an x-sequence. Accordingly, by equation (21),

$$\langle\!\langle \otimes^1, A'; 1\rangle\!\rangle \notin RF(D^l_{\beta*}). \tag{23}$$

Again, equations (22) and (23) contradict the fact that every minimal region of an E-maximal diagram has either shading or an x-sequence. Therefore, case (i) is impossible.

(ii) $\langle\!\langle \otimes^n, A_1; 1\rangle\!\rangle \in RF(D^j_\alpha)$, but $\langle\!\langle \otimes^n, A_2; 1\rangle\!\rangle \notin RF(D^l_{\beta*})$, where $\langle A_1, A_2\rangle \in \overline{cp}$.

Claim 2 of Theorem 22 says that $\langle\!\langle \text{Shading}, A_2; 1\rangle\!\rangle \notin RF(D^l_{\beta*})$, where $\langle A_1, A_2\rangle \in \overline{cp}$. The rest of the proof for this case is very similar to the proof for case (i). Case (ii) is also impossible.
Therefore,

$$\forall_{D^l_{\beta*}} \exists_{D^j_\gamma} (\{D^l_{\beta*}\} \models D^j_\gamma). \tag{24}$$

By equations (20), and (24) and Theorem 15,

$$\forall_{D^l_{\beta*}} \exists_{D^j_\gamma} (D^j_\gamma \sqsubseteq_{cp} D^l_{\beta*}). \tag{25}$$

By equation (25) and Theorem 9, we can transform every constituent of E-maximal diagram $D_{\beta*}$ to obtain an atomic diagram of D_α. That is,

$$\{D^1_{\beta*}\} \vdash D^{l_1}_\gamma, \quad \ldots, \quad \{D^m_{\beta*}\} \vdash D^{l_m}_\gamma. \tag{26}$$

By the rule of connecting a diagram and the rule of construction,

$$\{D^1_{\beta*} - \cdots - D^m_{\beta*}\} \vdash D^1_\gamma - \cdots - D^k_\gamma. \tag{27}$$

By equations (11), (12), (13), and (27),

$$\{D'_i\} \vdash D^1_\gamma - \cdots - D^k_\gamma. \tag{28}$$

Both D_j of (1) and each constituent of $D_\gamma^1 - \cdots - D_\gamma^k$ of (2) are parts of diagram D. Moreover, these exhaust all constituents of D. (Recall the assumption of Case 2). Therefore, by the rule of connecting a diagram, for every constituent of D', say, D_i', we get the following:

$$\{D_i'\} \vdash D. \tag{29}$$

By the rule of construction,

$$\{D'\} \vdash D. \tag{30}$$

Equations (8) (basis case) and (30) (inductive step) complete the proof. Therefore,

$$\{D'\} \vdash D. \tag{31}$$

By equations (2) and (31), $\Delta \vdash D$, which we wanted to show.

5
Venn-II and $\mathscr{L}_0$

In this chapter, I claim that Venn-II is equivalent to a first-order language $\mathscr{L}_0$, which I will specify in the first section. This claim is supported by two subclaims. One is that for every diagram D of Venn-II, there is a sentence φ of $\mathscr{L}_0$ such that the set assignments that satisfy D are isomorphic to the structures that satisfy φ. The other is that for every sentence φ of $\mathscr{L}_0$, there is a diagram D of Venn-II such that the structures that satisfy φ are isomorphic to the set assignments that satisfy D.

5.1 The language of $\mathscr{L}_0$

Our first-order language $\mathscr{L}_0$ is as follows:

A. Logical Symbols

(1) Parentheses: (,)
(2) Sentential connective symbols: $\neg, \wedge, \vee$
(3) Variables: $\mathbf{x}_1, \mathbf{x}_2, \mathbf{x}_3, \ldots$
(4) Equality: No

B. Parameters

(1) Quantifier symbols: $\forall, \exists$
(2) Predicate symbols: 1-place predicate symbols, $\mathbf{P}_1, \mathbf{P}_2, \ldots$
(3) Constant symbols: None
(4) Function symbols: None

5.2 From set assignments to structures

In this section, we want to show that there is an isomorphism between the set of set assignments for Venn-II and the set of structures for $\mathscr{L}_0$.

Because we have only one closed curve type and one rectangle, we need an extra mechanism in the semantics of this Venn system. That is a counterpart relation among tokens of a closed curve or among tokens of a rectangle. Before we make a mapping between sets and structures, we need to deal with these *cp*-related regions.

Let us name the basic regions enclosed by closed curves of well-formed diagrams, A_1, A_2,.... Given a *cp* relation, we can make equivalent sets on these basic regions: $[A_i] = \{x \mid \exists_{D\in\mathscr{D}}(x$ is a basic region of D and $A_i \equiv_{cp} x)\}$. Among the basic regions of each set $[A_i]$, choose the region whose name has the smallest index. (For example, if $A_2, A_4, A_8, \ldots$ are *cp*-related, then A_2 is the region for which we are looking.) We enumerate these chosen basic regions. Suppose we get the following: $A_{k_1}, \ldots, A_{k_n}, \ldots$; set assignment s assigns sets to these basic regions. Let $\mathscr{S}$ be the set of these set assignments. That is, when BR' is the set of the chosen basic regions (as just described) and U is a given domain,[1]

$$\mathscr{S} = \{s \mid s : BR' \longrightarrow \mathscr{P}(U)\}.$$

(N.B.: BR' is a subset of BR, i.e., the set of basic regions of well-formed diagrams.)

Let $\mathscr{M}$ be the set of structures for language $\mathscr{L}_0$. That is,

$$\mathscr{M} = \{m \mid m = \langle U, P_1^m, P_2^m, \ldots\rangle\}.$$

Then, there is an isomorphism h as follows:

$$h : \mathscr{S} \longrightarrow \mathscr{M}, \text{ where}$$

$$h(s_i) = \langle U, s_i(A_{k_1}), \ldots, s_i(A_{k_n}), \ldots\rangle, \text{ where}$$

$$s_i(A_{k_1}) = P_1^{m_i}, \ldots, s_i(A_{k_n}) = P_n^{m_i}, \ldots.$$

Let me introduce the following names for our further discussion:

$\mathscr{D}$ is the set of well-formed diagrams.
AD is the set of atomic well-formed diagrams.
BRG(*D*) is the set of basic regions of *D*.
RG(*D*) is the set of regions of *D*.
MRG(*D*) is the set of minimal regions of *D*.
RF(*D*) is the set of representing facts of *D*.

[1] I do not assume that U is not empty. If I assumed that U is not empty, I would need to introduce a rule of transformation such that we may put down an X-sequence in a diagram which consists only of a rectangle.

AF is the set of atomic formulas of $\mathscr{L}_0$, that is, $\{P_1x_1, P_2x_1, \ldots\}$.
FF is the smallest set of formulas to $\mathscr{L}_0$ satisfying the following:

(1) If α is in AF, then α is in FF.
(2) If α is in FF, then $(\neg\alpha)$ is in FF.
(3) If α and β are in FF, then $(\alpha \wedge \beta)$ and $(\alpha \vee \beta)$ are in FF.

SS is the set of sentences of $\mathscr{L}_0$.

What I will do in the next two sections is define the two following functions:

(1) $\overline{g} : \mathscr{D} \to SS$, where

$$\{h(s) \mid s \models D\} = \{m \mid m \models g(D)\}, \text{ where } D \in \mathscr{D}.$$

(2) $\overline{g'} : SS \to \mathscr{D}$, where

$$\{h^{-1}(m) \mid m \models \varphi\} = \{s \mid s \models \overline{g'}(\varphi)\}, \text{ where}$$

$\varphi \in \mathscr{L}_0$ and h^{-1} is the inverse of h.

5.3 From diagrams to sentences

In this section I aim to prove that for every diagram D of Venn-II there is a sentence φ of $\mathscr{L}_0$ such that the set assignments that satisfy D are isomorphic to the structures that satisfy φ.

First, we want to show how to get a sentence, say, φ, from a given diagram, say D. After getting φ, we will see that the isomorphism discussed in the previous section guarantees that the set of set assignments that satisfy D is isomorphic to the set of structures that satisfy φ. Let us start with an example:

Example 57 *Given the following diagram D, we want to get a sentence φ such that the structures that satisfy φ are isomorphic to the set assignments that satisfy D: Suppose that $\langle A_1, A_3\rangle \notin cp$, $\langle A_2, A_3\rangle \notin cp$.*

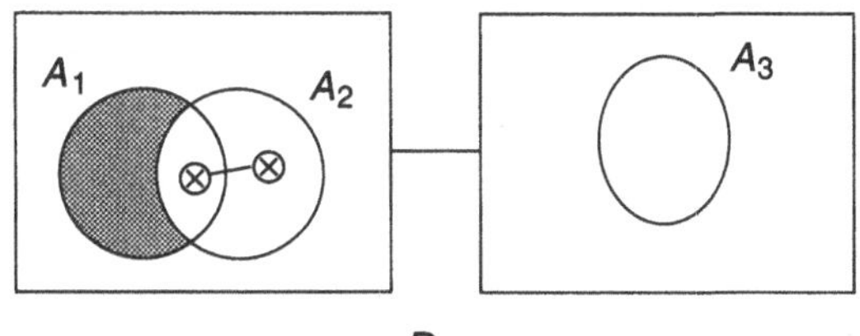

D

The first constituent of D has two representing facts: $\langle\langle$Shading, $A_1 - A_2; 1\rangle\rangle$ *and* $\langle\langle \otimes^2, (A_1\text{-and-}A_2) + (A_2 - A_1); 1\rangle\rangle$. *The second constituent of D does not have a representing fact. We want to get a sentence for each constituent of D and connect them by the disjunction symbol,* $\vee$. *If we assign a formula to a basic region, then we can assign a formula to each minimal region. Suppose basic region* A_1 *is assigned to formula* P_1x_1, A_2 *to* P_2x_1. *Then, minimal region* $A_1 - A_2$ *is assigned to formula* $(P_1x_1 \wedge \neg P_2x_1)$, A_1*-and-*A_2 *is to* $(P_1x_1 \wedge P_2x_1)$ *and* $A_2 - A_1$ *is to* $(P_2x_1 \wedge \neg P_1x_1)$. *Based upon these formulas and depending upon whether a minimal region is shaded or has an x-sequence, we get a sentence for each representing fact. Suppose we get a sentence* α_1 *for* $\langle\langle$*Shading,* $A_1 - A_2; 1\rangle\rangle$ *and* α_2 *for* $\langle\langle \otimes^2, (A_1\text{-and-}A_2) + (A_2 - A_1); 1\rangle\rangle$. *Then, we assign* $(\alpha_1 \wedge \alpha_2)$ *to the first atomic diagram of D. To the second atomic diagram, we will assign a logically true sentence of* $\mathscr{L}_0$*, say,* $\forall_{x_1}(P_1x_1 \rightarrow P_1x_1)$. *Therefore,* $\varphi = (\alpha_1 \wedge \alpha_2) \vee \forall_{x_1}(P_1x_1 \rightarrow P_1x_1)$. *After defining each function in detail, we will come back to this example to see how the functions work.*

Given diagram D and a set cp, this example suggests that we take three steps to find a sentence that is equivalent to D: First, assign formulas to regions. Second, get sentences out of these formulas depending upon whether the regions are shaded or have x-sequences. If we connect these sentences by $\wedge$ we get the sentence for each atomic diagram. However, if D is a diagram without any representing fact, that is, no shading or x-sequence, then we always get the same sentence, $\forall_{x_1}(P_1x_1 \rightarrow P_1x_1)$. Finally, if D is compound, we get a disjunctive sentence. Let us go through each step in detail.

5.3.1 From regions to formulas

Set $BRG(D)$, that is, set of basic regions of D, is a subset of set BR, that is, set of basic regions of well-formed diagrams. As described before, set BR' is a subset of set BR. Any basic region that is not in set BR' must have its cp-related region in BR'. Accordingly, for any basic region of D, say, R, R has its equivalent region in BR'. For example, if a region, say, R, is equivalent to A_{k_5} of set BR', then we assign formula P_5x_1 to this basic region R. In this way, equivalent regions get the same formula. We define function N as follows:

$N : BRG(D) \longrightarrow AF$, where

$$N(R) = P_ix_1 \text{ if } R \text{ is equivalent to } A_{k_i} \text{ in set } BR'.$$

Using this definition of function N, we want to assign a quantifier-free formula to a minimal region in the following way:

$N^0 : MRG(D) \longrightarrow FF$, where

> $MRG(D)$ is the set of minimal regions of *wfd* D, and $N^0(R^0) = (N(R_1^+) \wedge \cdots \wedge N(R_i^+)) \wedge \neg(N(R_1^-) \vee \cdots \vee N(R_k^-))$, where $R_1^+, \ldots, R_i^+$ are the basic regions (enclosed by closed curves) of which R^0 is a part, and $R_1^-, \ldots, R_k^-$ are the basic regions of which R^0 is not a part.[2]

5.3.2 From representing facts to sentences

Now we want to assign sentences to the representing facts of the diagram, using the formulas we got through function N^0. If D does not have any representing fact, that is, D has only a rectangle or closed curves, then we want to assign a tautological sentence.

For function f such that

$f : RF(D) \rightarrow SS$, where

$$f(\langle\langle \text{Shading},\ R_1^0 + \cdots + R_n^0; 1 \rangle\rangle) = \forall x_1(\neg N^0(R_1^0) \wedge \cdots \wedge \neg N^0(R_n^0))^3$$

and

$$f(\langle\langle \otimes^n,\ R_1^0 + \cdots + R_n^0; 1 \rangle\rangle) = \exists x_1(N^0(R_1^0) \vee \cdots \vee N^0(R_n^0)),^4$$

where $R_1^0, \ldots, R_n^0$ are minimal regions,

we can extend this function to get the mapping from $\mathscr{P}(RF(D))$ to SS in the following way:

$\bar{f} : \mathscr{P}(RF(D)) \rightarrow SS$, where

$$\bar{f}(\alpha) = \forall x_1(P_1 x_1 \rightarrow P_1 x_1) \text{ if} \alpha = \emptyset,$$
$$\bar{f}(\alpha) = \bigwedge f(\alpha_0) \text{ for each } \alpha_0 \in \alpha.^5$$

[2] We assume a certain order among the (negative) atomic conjuncts and among the (negative) atomic disjuncts. For example, we arrange each atomic formula in such a way that an atomic formula with a smaller index predicate comes before an atomic formula with a greater index predicate.

[3] We assume a certain order among the conjuncts.

[4] We assume a certain order among the disjuncts.

[5] We assume a certain order among the conjuncts.

5.3.3 From diagrams to sentences

If D is atomic, the sentence we got by function $\overline{f}$ is the sentence we are looking for. If D is compound, we make a disjunction of what we got for each constituent of D. That is, for function g such that $g : AD \rightarrow SS$, where $g(D) = \overline{f}(RF(D))$, we can extend this function to get the mapping from $\mathscr{D}$ to SS in the following way:

$\overline{g} : \mathscr{D} \rightarrow SS$, where

$$\overline{g}(D) = (g(D)) \text{ if } D \text{ is an atomic diagram,}$$
$$\overline{g}(D) = (\overline{g}(D_1) \vee \overline{g}(D_2)) \text{ if } D = D_1 - D_2.$$

Let us go back to Example 57 to see how function $\overline{g}$ works. Let us name the first constituent of D D_1 and the second one D_2. Then, the definition of function N says that $N(A_1) = P_1x_1$, $N(A_2) = P_2x_1$, and $N(A_3) = P_3x_1$. Before going through each step, let us figure out which sentence we want to map to this diagram in order that the diagram and the sentence are equivalent to each other. Intuitively, D_1 seems to be equivalent to sentence $(\forall x_1(P_1x_1 \rightarrow P_2x_1) \wedge \exists x_1 P_2x_1)$. Hence, we have to check whether or not function $\overline{g}$ gives us a sentence that is equivalent to $(\forall x_1(P_1x_1 \rightarrow P_2x_1) \wedge \exists x_1 P_2x_1)$.

$$\begin{aligned}
\overline{g}(D) &= (\overline{g}(D_1) \vee \overline{g}(D_2)). && (1)\\
\overline{g}(D_1) &= g(D_1) \\
&= \overline{f}(RF(D_1)) \\
&= f(\langle\!\langle \text{Shading}, A_1 - A_2; 1\rangle\!\rangle) \\
&\quad \wedge f(\langle\!\langle \otimes^2, (A_1\text{-}and\text{-}A_2) + (A_2 - A_1); 1\rangle\!\rangle) \\
&= \forall x_1 \neg N^0(A_1 - A_2) \wedge \exists x_1(N^0(A_1\text{-}and\text{-}A_2) \vee N^0(A_2 - A_1)) \\
&= \forall x_1(\neg(N(A_1) \wedge \neg N(A_2))) \\
&\quad \wedge \exists x_1((N(A_1) \wedge N(A_2)) \vee (N(A_2) \wedge \neg N(A_1))) \\
&= \forall x_1(\neg P_1x_1 \vee P_2x_1) \\
&\quad \wedge \exists x_1((P_1x_1 \wedge P_2x_1) \vee (P_2x_1 \wedge \neg P_1x_1)). && (2)\\
\overline{g}(D_2) &= (\forall x_1(P_1x_1 \rightarrow P_1x_1)). && (3)
\end{aligned}$$

By equations (1), (2) and (3),

$$\begin{aligned}
\overline{g}(D) &= (\forall x_1(\neg P_1x_1 \vee P_2x_1) \wedge \exists x_1((P_1x_1 \wedge P_2x_1) \\
&\quad \vee (P_2x_1 \wedge \neg P_1x_1))) \vee (\forall x_1(P_1x_1 \rightarrow P_1x_1)).
\end{aligned}$$

This sentence is equivalent to sentence $(\forall x_1(P_1x_1 \rightarrow P_2x_1) \wedge \exists x_1 P_2x_1) \vee (\forall x_1(P_1x_1 \rightarrow P_1x_1))$. Now, the question is whether the set assignments that satisfy D are isomorphic to the structures that satisfy this sentence. However, the isomorphism h defined in §5.2 tells us that they are.

Claim I *Let h be the function defined in §5.2 and $\overline{g}$ be as above. Then, for any diagram D, $\{h(s) \mid s \models D\} = \{m \mid m \models \overline{g}(D)\}$.*

Proof It follows by induction on the length of diagram D.

(Basis case) Let D be an atomic diagram. Let $s \models D$. That is, $\forall_{\alpha \in RF(D)}(s \models \alpha)$. There are two cases:

(i) $\alpha = \langle\langle \text{Shading}, R; 1\rangle\rangle$. That is, $s \models \langle\langle \text{Shading}, R; 1\rangle\rangle$. (We want to show that $\{h(s) \mid s \models \langle\langle \text{Shading}, R; 1\rangle\rangle\} = \{m \mid m \models f(\langle\langle \text{Shading}, R; 1\rangle\rangle)\}$. I.e., $\{h(s) \mid s \models \langle\langle \text{Shading}, R; 1\rangle\rangle\} = \{m \mid m \models \forall x_1(\neg N^0(R_1^0) \wedge \cdots \wedge \neg N^0(R_n^0))\}$.) We prove this by induction on the number of minimal regions that comprise region R. Suppose that R itself is a minimal region. Let $R = (R_1^+\text{-}and\text{-}\cdots\text{-}and\text{-}R_i^+) - (R_1^- + \cdots + R_j^-)$. Hence, $\overline{s}(R) = (s(R_1^+) \cap \cdots \cap s(R_i^+)) - (s(R_1^-) \cup \cdots \cup s(R_j^+))$. $N^0(R) = (N(R_1^+) \wedge \cdots \wedge N(R_i^+)) \wedge \neg(N(R_1^- \vee \cdots \vee N(R_j^-))$. Therefore, by the isomorphism h defined in §5.2, $\{h(s) \mid s \models \langle\langle \text{Shading}, R; 1\rangle\rangle\} = \{m \mid m \models \forall x_1 \neg N^0(R)\}$. (I leave the case in which R consists of n minimal regions to the reader.)

(ii) $\alpha = \langle\langle \otimes^n, R; 1\rangle\rangle$. This is very similar to (i).

(Inductive case) Suppose that $D = D_1 - D_2$. By the inductive hypothesis, $\{h(s) \mid s \models D_1\} = \{m \mid m \models \overline{g}(D_1)\}$ and $\{h(s) \mid s \models D_2\} = \{m \mid m \models \overline{g}(D_2)\}$. Since $\overline{g}(D) = (\overline{g}(D_1) \vee \overline{g}(D_2))$, obviously, $\{h(s) \mid s \models D\} = \{m \mid m \models \overline{g}(D)\}$. □

5.4 From sentences to diagrams

In this section I aim to prove that for every sentence φ of $\mathscr{L}_0$ there is a diagram D of Venn-II such that the structures that satisfy φ are isomorphic to the set assignments that satisfy D.

Let me introduce some terminology for further discussion. We call sentence β **quantifier-literal** (simply, *Q-literal*) if $\beta = \forall x \bigvee \alpha_i$ or $\beta = \exists x \bigwedge \alpha_i$, and α_i are atomic formulas or the negation of atomic formulas. We say sentence φ is in **quantifier disjunctive normal form** (simply, *Q-DNF*) if $\varphi = \bigvee \alpha_i$, $\alpha_i = \bigwedge \beta_j$ and β_j are Q-literals.

Theorem 24 *For any sentence φ' of $\mathscr{L}_0$, there is a sentence φ such that φ is in Q-DNF and φ and φ' are logically equivalent to each other.*

Proof We transform sentence φ' into a sentence in prenex normal form, say, φ''. That is, $\varphi'' = Q_1 x_1 \cdots Q_n x_n \alpha$, where α is a quantifier-free formula.

(i) If Q_n is a universal quantifier, then transform α into a formula in conjunctive normal form. That is, $\alpha = \alpha_1 \wedge \cdots \wedge \alpha_m$. Distribute "$\forall x_n$" over "$\wedge$." We get $Q_1 x_1 \cdots Q_{n-1} x_{n-1}(\forall x_n \alpha_1 \wedge \cdots \wedge \forall x_n \alpha_m)$. Each conjunct of $\forall x_n \alpha_1 \wedge \cdots \wedge \forall x_n \alpha_m$ should have the following form: $\forall x_n(\beta_1 \vee \cdots \vee \beta_k)$, where each $\beta_i (1 \leq i \leq k)$ is an atomic formula or the negation of an atomic formula. Now we want to unpack formula $\forall x_n(\beta_1 \vee \cdots \vee \beta_k)$ in the following way: If β_i does not have variable x_n, then remove β_i from the scope of $\forall x_n$. (We know that $\forall x_n \beta_i$ is equivalent to β_i if β_i does not have variable x_n.) Hence, we get formula $\forall x_n(\beta_{i_1} \vee \cdots \vee \beta_{i_l}) \vee (\beta_{i_{l+1}} \vee \cdots \vee \beta_{i_k})$, where $\beta_{i_1}, \ldots, \beta_{i_l}$ are the formulas with variable x_n and $\beta_{i_{l+1}}, \ldots, \beta_{i_k}$ are the formulas that do not have variable x_n. Since $\mathscr{L}_0$ is a monadic language, we know that formula $\forall x_n(\beta_{i_1} \vee \cdots \vee \beta_{i_l})$ does not have any free variable. Therefore, we treat this formula as if it were atomic in the remaining process.

(ii) If Q_n is an existential quantifier, then transform α into a formula in disjunctive normal form. That is, $\alpha = \alpha_1 \vee \ldots \vee \alpha_m$. Perform steps similar to those in (i).

We repeat either (i) or (ii) depending upon whether the quantifier is universal or existential, until we distribute the last quantifier, $Q_1 x_1$. If Q_1 is universal, then the final sentence is what we want. If Q_1 is an existential quantifier, we get the sentence in Q-CNF. In this case, we change this sentence to a sentence in Q-DNF. □

Since every sentence can be transformed into a logically equivalent sentence in Q-DNF, we want to provide a way to get a diagram for a sentence in Q-DNF. A sentence in Q-DNF is a disjunction of one or more sentences, each of which is a conjunction of one or more Q-literals. If we can construct a mapping from Q-literals to diagrams, then a conjunction of Q-literals corresponds to the unification of the diagrams. A disjunction of these conjunctive Q-literals corresponds to connecting the diagrams.

Example 58 *Let us start with the sentence used in Example 57,* $(\forall x_1(P_1 x_1 \rightarrow P_2 x_1) \wedge \exists x_1 P_2 x_1) \vee \forall x_1(P_1 x_1 \rightarrow P_1 x_1)$. *This sentence is in Q-DNF, which consists of two disjuncts. We need to get one atomic diagram for each disjunct. The first disjunct consists of two conjuncts. This means that the atomic diagram for this disjunct will have two representing facts. Obviously,* *we want to draw a shading for a universal sentence and an x-sequence for an existential sentence.*

This example suggests that we need four steps to find a diagram for a given sentence: First, we map a formula to a region. Second, we assign a representing fact, that is, whether the region is shaded or has an X-sequence, to a Q-literal, depending on whether the Q-literal is a universal sentence or an existential sentence. Hence, we get a diagram for each Q-literal. Third, we get a diagram for the conjunction of Q-literals by unifying the diagrams of the Q-literals. Fourth, we connect the diagrams, each of which is a diagram for a disjunct of the given sentence in Q-DNF. Let us formalize these four steps in the following subsections.

5.4.1 From formulas to regions

We want to assign region A_{k_i} of BR' to atomic formula $P_i x_1$ of set AF. We can extend this relation to the relation between set FF and the set of regions as follows:

For function N' such that N': $AF \to BR'$, where $N'(P_n x_i) = A_{k_n}$, we can extend this function to get the mapping from FF to RG in the following way:

$$\overline{N'} : FF \to RG,$$

where

$$\begin{aligned}
\overline{N'}(\alpha) &= N'(\alpha) \quad \text{if } \alpha \in AF, \\
\overline{N'}(\alpha) &= R - \overline{N'}(\beta), \text{ where } R \text{ is a rectangle,} \quad \text{if } \alpha = (\neg\beta), \\
\overline{N'}(\alpha) &= \overline{N'}(\alpha_1)\text{-}\mathit{and}\text{-}\overline{N'}(\alpha_2) \quad \text{if} \alpha = (\alpha_1 \wedge \alpha_2), \\
\overline{N'}(\alpha) &= \overline{N'}(\alpha_1) + \overline{N'}(\alpha_2) \quad \text{if} \alpha = (\alpha_1 \vee \alpha_2).
\end{aligned}$$

5.4.2 From sentences to diagrams

We want to assign an atomic diagram to a Q-literal. In the previous subsection, we assigned the same formula to any atomic diagram with no representing fact. Here we want to assign the diagram that consists only of a rectangle to any tautological Q-literal. We also assign to a contradictory Q-literal the diagram that consists only of a rectangle both with a shading and with an x-sequence. That is,

Let *Q-Lit* be the set of Q-literals and R_0 be a rectangle. For any function g' such that g': *Q-Lit* $\rightarrow AD$, where φ is a logically true Q-literal, then $g'(\varphi) = D$, with

$$Seq(D) = \langle \{R_0\},\ \emptyset,\ \emptyset \rangle,$$ [6]

if φ is a contradiction, then $g'(\varphi) = D$, with

$$Seq(D) = \langle \{R_0\},\ \{R_0\},\ \{R_0\} \rangle,$$

and if φ is neither a tautology nor a contradiction, then $g'(\forall x_n \alpha) = D$, with

$$Seq(D) = \langle \{N'(\alpha_i) \mid \alpha_i \text{ is an atomic formula in } \alpha\} \cup \{R_0\},\ \{\overline{N'}(\neg\alpha)\},\ \emptyset \rangle,$$

and $g'(\exists x_n \alpha) = D$, with

$$Seq(D) = \langle \{N'(\alpha_i) \mid \alpha_i \text{ is an atomic formula in } \alpha\} \cup \{R_0\},\ \emptyset,\ \{\overline{N'}(\alpha)\} \rangle,$$

we can extend this function to get the mapping from the set of sentences in Q-DNF to $\mathscr{D}$ in the following way: Let *Q-DNFs* be the set of sentences in Q-DNF.

$\overline{g'}$: *Q-DNFs* $\rightarrow \mathscr{D}$, where

$$\begin{aligned} \overline{g'}(\alpha) &= g'(\alpha) \quad \text{if } \alpha \in Q\text{-}Lit, \\ \overline{g'}(\alpha) &= D, \text{ if } \alpha = (\alpha_1 \wedge \alpha_2), \end{aligned}$$

where $\{\overline{g'}(\alpha_1), \overline{g'}(\alpha_2)\} \rightsquigarrow D$ by the unification rule,

$$\overline{g'}(\alpha) = D, \text{ if } \alpha = (\alpha_1 \vee \alpha_2),$$

where $D = \overline{g'}(\alpha_1)\text{–}\overline{g'}(\alpha_2)$.

Let us go back to Example 58 to see how the function $\overline{g'}$ works.

$$\begin{aligned} &\overline{g'}((\forall x_1(P_1x_1 \rightarrow P_2x_1) \wedge \exists x_1 P_2x_1) \vee \forall x_1(P_1x_1 \rightarrow P_1x_1)) \\ &\quad = \overline{g'}(\forall x_1(P_1x_1 \rightarrow P_2x_1) \wedge \exists x_1 P_2x_1)\text{–}\overline{g'}(\forall x_1(P_1x_1 \rightarrow P_1x_1)) \\ &\quad = D'\text{–}\overline{g'}(\forall x_1(P_1x_1 \rightarrow P_1x_1)), \end{aligned}$$

where $\{\overline{g'}(\forall x_1(P_1x_1 \rightarrow P_2x_1)), \overline{g'}(\exists x_1 P_2x_1)\} \rightsquigarrow D'$ by the unification rule.

$$\overline{g'}(\forall x_1(P_1x_1 \rightarrow P_2x_1)) = \overline{g'}(\forall x_1(\neg P_1x_1 \vee P_2x_1)) = D_1,$$

[6] Recall the definition of the sequence of a *wfd*. This shows that D is atomic, D has the first element of this sequence as the set of its basic regions, the second element as the shaded regions, and the third one as the regions with x-sequences.

where

$$Seq(D_1) = \langle \{R_0, A_1, A_2\}, \{A_1 - A_2\}, \emptyset \rangle.$$

$$\overline{g'}(\exists x_1 P_2 x_1) = D_2,$$

where

$$Seq(D_2) = \langle \{R_0, A_2\}, \emptyset, \{A_2\} \rangle.$$

Unify these two diagrams to get, say, D'. Then, $Seq(D') = \langle \{R_0, A_1, A_2\}, \{A_1 - A_2\}, \{A_2\} \rangle$.

$$\overline{g'}(\forall x_1 (\neg P_1 x_1 \vee P_1 x_1)) = D'',$$

where $Seq(D'') = \langle \{R_0\}, \emptyset, \emptyset \rangle$.

Now we connect D' and D'' as follows:

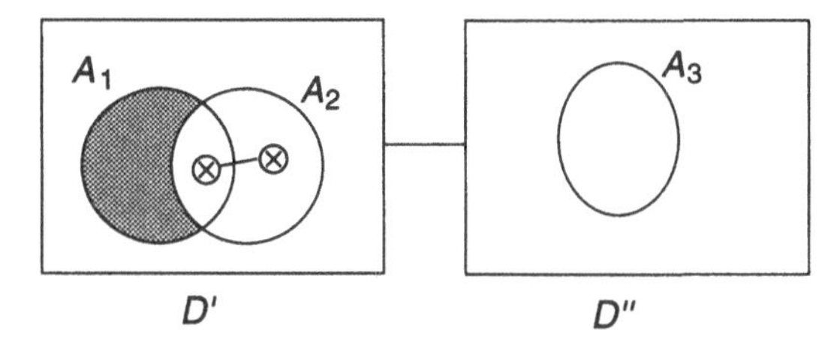

Claim II *Let h^{-1} be the inverse function of h defined in §5.2 and $\overline{g'}$ be as before. Then, for any sentence φ, $\{h^{-1}(m) \mid m \models \varphi\} = \{s \mid s \models \overline{g'}(\varphi)\}$*

Proof First, we transform sentence φ into a sentence in Q-DNF that is equivalent to φ. Let us call it φ'. Now, we prove this claim by induction on φ'.

(Basis case) There are four cases: φ' is logically true, a contradiction, $\exists x_n \alpha$, or $\forall x_n \alpha$. In the case of a tautology or a contradiction, this claim is trivially true. Suppose that $\varphi' = \exists x_n \alpha$. The case where $\varphi' = \forall x_n \alpha$ is similar. That is, $\varphi' = \exists x_n (\neg) \bigwedge P_i x_n$. We want to show $\{h^{-1}(m) \mid m \models \exists x_n (\neg) \bigwedge P_i x_n\} = \{s \mid s \models \overline{g'}(\exists x_n (\neg) \bigwedge P_i x_n)\}$. We can show it by induction on the number of conjuncts of "$(\neg) \bigwedge P_i x_n$." We will check one of the basis cases. (I will leave the inductive step to the reader, which is straightforward.) There are two basis cases: when $\varphi' = \exists x_n P_i x_n$ or when $\varphi' = \exists x_n \neg P_i x_n$. Suppose that $\varphi' = \exists x_n P_i x_n$. Function $\overline{g'}$ assigns to this sentence a well-formed diagram with a rectangle and a closed curve whose name is $\overline{N'}(P_i)$, that is, A_i. Also, this diagram has one representing fact: $\langle\langle \otimes^1, A_1; 1 \rangle\rangle$. Let call this diagram D_α. Therefore,

$\{h^{-1}(m) \mid m \models \exists x_n P_i x_n\} = \{s \mid s \models D_\alpha\}$. The other case, that is, when $\varphi' = \exists x_n \neg P_i x_n$, is very similar.

(Inductive case) There are two inductive cases: when $\varphi' = (\varphi_1 \wedge \varphi_2)$ or when $\varphi' = (\varphi_1 \vee \varphi_2)$. In each case, the inductive hypothesis yields us easily what we want to prove. □

Claim I at the end of §5.3.3 and Claim II at the end of §5.4.2 show that Venn-II and the first-order language $\mathscr{L}_0$ are equivalent to each other.

6
Diagrammatic versus Linguistic Representation

I started this work with the following assumption: There is a distinction between diagrammatic and linguistic representation, and Venn diagrams are a nonlinguistic form of representation. By showing that the Venn system is sound and complete, I aimed to provide a legitimate reason why logicians who care about valid reasoning should be interested in nonlinguistic representation systems as well. However, the following objection might undermine the import of my project: How do we know that Venn diagrammatic representation is a nonlinguistic system? After all, it might be another linguistic representation system which is too restricted in its expressiveness to be used in our reasoning. If we accept this criticism, what I have done so far would be reduced to the following: A very limited linguistic system was chosen and proven to be sound and complete. Considering how far symbolic logic has developed, this could not be an interesting or meaningful project at all. Therefore, it seems very important to clarify the assumptions that diagrammatic systems are different from linguistic ones and that the Venn systems are nonlinguistic.

There has been some controversy over how to define diagrams in general, despite the fact that we all seem to have some intuitive understanding of diagrams. For example, all of us seem to make some distinction between Venn diagrams and first-order languages. Everyone would classify Euler circles under the same category as Venn diagrams. Or suppose that both a map and verbal instructions are available for us to locate a certain place. None of us would think that these two media are of the same kind. But, in some other cases, our intuition does not guide us one way or another. For example, is a timetable with lines and English words diagrammatic or linguistic? Is a pie chart diagrammatic, in spite of the figures and English words in each section?

In this chapter, I aim to reexamine the issues involved in diagrammatic versus linguistic representation, and to justify the assumption that Venn systems are a nonlinguistic form of representation.

In the first section, I will review two main criteria that have been suggested by many people to distinguish between diagrams and linguistic texts. These unsuccessful distinctions lead us to the following question about the necessity of the classification of these two kinds of systems: Is there any other motive behind this classification besides theoretical accuracy? This question turns my interest from the definition of diagrams in general to finding basic differences between diagrammatic and linguistic features.

In the second section, I will sharpen our intuition behind a distinction between these two kinds of systems. It has been said that a picture depicts reality by its resemblance to it whereas a sentence represents it by convention. Diagrams are believed to be closer to pictures in this sense. One of the assumptions behind this intuition is that resemblance and convention are complementary. I will argue that conventionality is not inversely proportional to resemblance to reality, but to perceptual inferences. That is, the less a system relies on our perceptual inferences, the more conventions are introduced. Based on the reliance on perceptual inferences (or, inversely, on conventions) I will present three aspects of representation in which diagrammatic and linguistic features are different from each other.

In the third section, I will show how these three characteristics concerning perceptual inferences are used in the Venn systems. This examination will support the implicit assumption of the present work that the presented Venn systems are diagrammatic forms of representation.

In the fourth section, a criticism of diagrammatic features of representation will be discussed. Certain kinds of information, that is, disjunctive information or negative information, are not easily represented perceptually. When a system aims to handle these kinds of reasoning, it needs to bring linguistic features into the system. I will point out an aspect of the Venn systems that can be classified as a linguistic feature.

6.1 Visuality and geometrical figures

Let us start with the following systems most of which we have discussed:

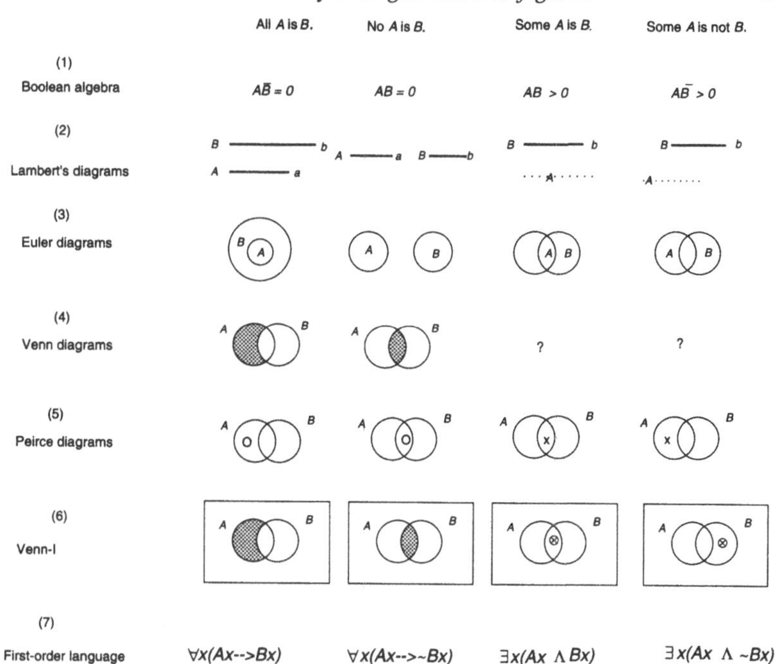

Many of us would not hesitate to call both Boolean algebra and a first-order language linguistic, and the rest of them diagrammatic. This classification has been accepted for at least the two following reasons: First, the systems in (2) through (6) are visual in that they try to represent relations between classes more visually, compared with (1) and (7). Second, the systems (2)–(6) adopt geometrical figures (in addition to letters) to represent classes. However, I will show in this section that these two criteria are not sufficient to be accepted without controversy.

As for the first criterion, that is, visuality, the meaning of "visual" is not clear, and the word "visual" could be misleading. Someone might say that sentences in a first-order language written on a board are visual in the sense that we can see them. Someone might say that Venn diagrams in braille are not visual in the sense that we should touch (as opposed to see) them. In this sense, "visual" and "linguistic" seem to be compatible. According to this ordinary sense of "visual," it would not be correct to call first-order languages linguistic and the Venn systems visual. Also, I find in related literature that "diagrammatic" and "visual" are used synonymously. In this case, it is very important to make it clear that the use of the word "visual" is different from the ordinary use of this word. We also need to clarify the meaning of "being visual" before we call some representation system diagrammatic because of its visuality.

As for the second criterion, that is, the nature of the primitive objects of the system, we can ask whether a representation system is diagrammatic if and only if the system adopts geometrical figures. There seems no difficulty in imagining a first-order language with some geometrical figures as its predicates or constants. On the other hand, there are some representation systems that we might want to call diagrammatic despite the fact that they consist more of letters than of geometrical figures. Also, the distinction between geometrical figures and symbols does not seem to be clear either. Let me elaborate this point in detail in the following simple example which does not involve any diagrammatic system we have discussed.

Suppose Tom, Susan, and Mary are sitting on three chairs in a row. Suppose we want to represent the following situation to our friend who is curious to know who sits next to whom: Tom is to the right of Susan and Susan is to the right of Mary.

(Case 1) We may take a photograph of this situation and show it to him.

(Case 2) We may draw the following diagram: Let "□" denote Tom, "○" Susan, and "△" Mary.

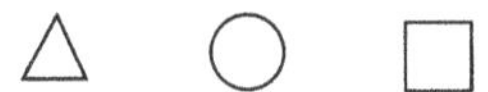

(Case 3) We may write the following first-order sentence:

$$RightOf(t, s) \wedge RightOf(s, m),$$

where "t" denotes Tom, "s" Susan, and "m" Mary.

Even though we are concerned with the same aspect of the same situation and all of these three cases succeed in addressing the concern in question, there seems to be a clear difference among these media. None of us would confuse a photograph with a diagram, a diagram with a sentence, or a photograph with a sentence.

The first thing we notice is that each medium consists of different kinds of objects. A photograph consists of images produced by a camera, a diagram has geometrical figures and a sentence consists of symbols. This obvious noticeable fact might have led us to the second criterion we mentioned previously. That is, the property of primitive objects adopted in a system determines whether the system is diagrammatic or not. Can we represent this information diagrammatically, but without a geometrical figure? Or, can we come up with a linguistic form of representation with geometrical figures? Let us consider the following two cases:

(Case 4)

Mary Susan Tom

(Case 5)

$$RightOf(\square, \bigcirc) \wedge RightOf(\bigcirc, \triangle),$$

where "□" denotes Tom, "○" Susan, and "△" Mary.

Many classify case 4 as a diagrammatic representation along with case 2, while classifying case 5 as a linguistic one together with case 3. Despite the fact that we used letters of the alphabet, most of us have a strong tendency to call case 4 diagrammatic rather than linguistic. Also, using geometrical figures does not prevent us from saying that case 5 is a sentence of some first-order language, if this language contains "□", "○", "△" as its constants. If this intuition is correct, the previous biconditional – A representation system is diagrammatic if and only if the system adopts geometrical figures – is not true in either direction.

This discussion shows that we seem to have an intuition, albeit naive, that classifies photographs, diagrams, and linguistic texts. However, this discussion assures us that the type of primitive objects of a representation system does not determine the type of the representation system. In this sense, the fact that the primitive objects of Venn-I and Venn-II, that is, closed curves, shading, lines, and so forth, are geometrical figures (not linguistic symbols) should not convince us that these systems are diagrammatic or nonlinguistic.

6.2 Perceptual inferences

In this section, I examine another suggestion that has been in the air. I will modify and sharpen the ideas behind this suggestion. The suggestion is that the distinction among photographs, diagrams, and linguistic texts depends upon the degree of conventionality. We do not need to learn any conventions for a photograph, and we need fewer conventions for a diagram than for a first-order language. For example, if we know what Tom, Susan, and Mary look like, there should be no problem in understanding the situation by looking at the photograph of case 1. If we are told that a rectangle denotes Tom, a circle Susan, and a triangle Mary, we can easily get the information we want to know from the diagram of case 2. However, in order to understand the sentence of case 3, we have to understand a meaningful unit of this system, that is, $RightOf(a, b)$, the meaning of a logical symbol, $\wedge$, and so on. The picture of case 1 is thought to be more similar to reality than the diagram of case 2 or the sentence of case 3, and the diagram of case 2 more similar to reality than the sentence of case 3. The more similar to reality a mode of representation is, the less conventional it is. This suggestion is related to one of our naive distinctions between pictures and words, which says that words denote objects by convention and pictures or diagrams by similarity. This criterion seems to work for our additional two cases, cases 4 and 5.[1] As far as the degree of convention goes, cases 3 and 5 require one to learn more conventions than cases 2 and 4.

There are three problems we have to be aware of with this intuition. First, the meaning of the similarity by which pictures refer to objects should be clarified. The rectangle by which we refer to Tom in case 2 is no more similar to Tom than the word "Tom" is to Tom. What is worse, if we aim to represent abstract reality, we cannot talk about similarity. If a diagram relied on similarity only, then it would be impossible for us to have a diagrammatic form of representation for these abstract realities. Second, this distinction seems to assume that a diagrammatic system adopts pictures and a linguistic system words. As we have seen, some representations seem to be diagrammatic even with words (as in case 4) and vice versa (as in case 5). Third, if we can talk about the degree of similarity (or conventionality), then does this mean that we can talk about the degree of diagrammaticality (or linguisticality)? However, this naive intuition, relying on similarity and convention, I think, carries a point that it is worthwhile to consider. I claim these three problems can be solved simultaneously if we clarify the sources of our intuition about "similarity to reality."

It is often said that a picture is quite similar to the reality of which it is a picture. What do we mean by "similar to reality?" I am aware that complicated discussions are needed for the relation "being similar" between a picture and and the reality. First of all, a picture is two-dimensional, whereas reality is three-dimensional. An image in a picture does not carry the texture we can feel from real objects, and so on. What makes these two different kinds of objects similar to each other? I do

[1]Case 4 is a diagrammatic representation, but with the names of the people in the letters. Case 5 includes geometrical figures in a sentence, but is still a linguistic text.

not intend to become involved in answering this question. Rather, I accept the following intuition: An image of Tom in a picture (as in case 1) is more similar to the person Tom in reality than a rectangle (as in case 2) that denotes Tom. I accept this intuition in the following sense: It is more *obvious* which person is represented by looking at a picture than by seeing a rectangle. Pictures rely on our perceptual abilities more than diagrams or symbols do. We need to learn fewer conventions for pictures than for diagrams or symbols. Looking at a photograph and understanding the reality of which it is a picture do not require much intellectual effort.

What makes a picture of Tom similar to Tom is beyond the scope of the present work. However, the following seems to be quite obvious: The fact that a picture relies on our perceptual abilities more than a diagram or symbol does is closely related to our intuition that a picture is more similar to reality than a diagram or symbol is. I will explore this point further and claim that the same kind of relation holds between diagrams and linguistic texts. That is, diagrams rely on our perceptual inferences more than linguistic texts do. Accordingly, linguistic texts use more conventions than diagrams do.

We know that a photographic style of representation does not get us too far. In many cases, we want to represent what we cannot take a picture of. For example, we want to represent sets, relations of sets, love–hate relations of people, family relations, and so on. In order to reason about these things by means of representation systems, we need to invent some conventions to represent these entities, whether it is a diagrammatic representation system or a linguistic representation system.

More specifically, let us compare Venn-I and a first-order language of set theory. Both systems aim to represent sets. Since we do not know what sets, even if they exist in some sense, look like, it does not make sense to talk about the similarity between representing objects (the primitive objects of Venn-I or of a set-theoretic language) and represented objects (sets). If we adhere to the distinction between diagrammatic and linguistic systems that the former denotes objects by similarity and the latter by convention, then we should conclude that there can be no diagrammatic representation of sets. In a language of set theory, letters $a, b, \ldots,$ might be adopted to denote sets. Similarly, circles are adopted in the Venn systems. Each of them has its own convention. Neither of them relies on our perceptual inferences between representing objects and represented objects. Accordingly, the condition that primitive objects of a diagrammatic system should be similar to what they represent is too restrictive. This simple example is enough to convince us that the meaning of "similarity to reality" should be examined in different aspects.

I examine the differences in degrees of perceptual inferences between diagrammatic and linguistic systems from the following three angles: the way relations among objects are represented, the way conjunctive information is represented, and the way tautological or contradictory information is represented.

6.2.1 Relations among objects

A representation system seeks to represent some information by means of its own media. In order to achieve this goal, in most cases a system needs to represent not only objects but relations among them. For example, when the Venn systems aim to represent the information "Every unicorn is red" in terms of the relation of sets, these systems need to represent relations of sets as well as sets. In the case of linguistic representation systems, some arbitrary symbols which do not bring ambiguity into the systems are chosen for the representation of sets and relations of sets. On the other hand, we have thought that Venn diagrammatic systems represent these concepts by *visualizing* them in one way or another. However, we have already noted the controversy over the meaning of visualization.

Let us go back to the previous examples of case 2 and case 3. In order to convey information about the position of the three people, three people and the relations among them should be represented. For the three people, each case adopted its own convention: Case 2 adopts geometrical figures, that is, box, circle, and triangle, and case 3 letters, t, s, and m. Up to this point, there is no intrinsic difference between case 2 and case 3.

For the relation of "being to the right of," in case 3, which is a linguistic representation, the symbol "*RightOf*" was chosen just as symbol t is for Tom. This is an arbitrary choice since it could be "*Right*" or "*RO*" or almost anything else. This symbol brings in its own convention. For example, "$RightOf(t)$" is ungrammatical, $RightOf(t,s)$ is different from $RightOf(s,t)$, and so on. On the other hand, in case 2, which is a diagrammatic representation, no new object was introduced for this concept. Instead, the objects that represent the individuals are arranged in a way very similar to the way the three people are sitting in reality. There is no convention to learn in this case, unlike with case 3.

This is one of the ways in which we can extend the meaning of similarity between diagrammatic systems and reality. Suppose we want to represent a relation among objects. If this is a spatial relation, a diagram can represent it without any additional symbol or convention. The relation "being to the right of" is a good example of a spatial relation. An isomorphism between a spatial relation that we want to represent and a spatial arrangement in a diagram is perceptually so obvious that there is no need for any extra syntactic device or convention for this relation. This perceptual obviousness of relations is parallel to what we saw earlier with photographs and individuals. An isomorphism between an image of Tom on a picture and Tom is perceptually so obvious that we say that a picture is more similar to reality than a diagram or a symbol is. Larkin and Simon note the parallelism in the case of diagrams:

> The fundamental difference between our diagrammatic and sentential representations is that the diagrammatic representation preserves explicitly the information about the topological and geometrical relations among the components of the problem, ...[2]

This is one of the main reasons why we prefer maps to verbal instructions in order to locate a place. For this purpose, spatial relations among places are the most significant. These spatial relations are represented isomorphically in diagrams in terms of the spatial arrangements of the representing objects. Maps do not need any new syntactic device for the relations among the places represented by the objects in the map. If one place is north of another place, then this spatial relation is mirrored in a map in a perceptually obvious way.

Suppose the relation we want to represent is not a spatial one. Suppose Tom is a member of the set of American dentists. Tom's membership in the set of American dentists is not a spatial relation, whereas the relation "being to the right of" is. Let us recall the case in which we aim to represent abstract entities. Diagrammatic representations also need to come up with artificial syntactic devices to represent them. If a relation is not spatial, do we need to come up with a new syntactic device? Even in this case, diagrammatic systems do not bring in convention, but represent a nonspatial relation in terms of a spatial arrangement among objects. Accordingly, no convention need be learned. For example, we want to represent the relation "being member of." Suppose that a system has a dot to represent an individual and a box a set. In order to represent the relation "being a member of" that holds between an individual and a set, most probably this system requires us to put down

[2]Larkin and Simon [16], p. 66.

a dot *in* a box.[3] Putting down an object *in* another object does not involve any new syntactic device. A spatial arrangement between a dot and a box represents a nonspatial relation, "being a member of." An isomorphism between this spatial arrangement and the relation "being a member of" is not as perceptually obvious as an isomorphism between a spatial arrangement in a diagram (as in case 2) and the relation "being to the right of." However, this isomorphism is more perceptually obvious than any linguistic symbol that a linguistic representation adopts, since no extra convention involving syntactic devices is required.

Timetables might be a good example to illustrate this point. What we aim to represent in a timetable is the temporal relations (not spatial relations) among the events. A timetable translates this temporal relation into a spatial one, either by placing one appointment above another or by placing one to the right of another. Family trees which are used for the relations among a family reveal this diagrammatic feature quite well. Family relations are not spatial. What we are interested in is certain kinds of relations among family members, for example, who is the father of whom, who is the uncle of whom, and so on. In a family tree these relations are transformed into spatial relations.

So far, I have argued that diagrammatic systems need fewer conventions in representing relations than linguistic systems do. In the case of linguistic systems, any relation is represented by some symbolic devices. These new syntactic devices are accompanied by their own conventions which we need to learn in order to understand the systems. On the other hand, diagrammatic systems tend to represent relations in terms of spatial arrangements among objects, not in terms of new devices. If a relation is spatial, then the representation in a diagram is perceptually obvious. The relations "being to the left of," "being to the right of," "being in front of," and "being behind" are examples of spatial relations. Even if a relation is not spatial, some relations are representable by means of spatial relations among objects. We have discussed the relation "being a member of" as an example. In many cases, however, relations are not easily representable in terms of spatial arrangements. For example, the relation "love" is not easy to represent in terms of a spatial relation between objects, unlike "being a member of." In this case, most probably some syntactic device should be introduced in diagrammatic systems. Then, what makes one system diagrammatic and another linguistic if conventions are needed in both cases? I suggest another

[3]This will be explained more fully in §6.3.1.

extended way to interpret the similarity relation between diagrams and reality in the following subsection. This similarity leads to an obvious perceptual inference.

6.2.2 Conjunctive information

Most representation systems are designed to represent more than one piece of information. In this subsection, I will claim that there is a fundamental difference between diagrammatic and linguistic representations in the way more than one piece of information is represented. First, I will show that the way conjunctive information is represented in diagrams is more similar to reality than it is in linguistic texts. Also, the way a diagrammatic system represents conjunctive information heavily relies on our perceptual abilities, rather than on conventions about syntactic devices. Accordingly, fewer conventions are involved in the former than in the latter.

Suppose there are chairs in a row. Tom is sitting on one of the chairs. Susan is told to sit to the right of Tom from our point of view. After Susan takes a seat, Mary is told to sit to the right of Susan. As soon as Mary seats herself, the fact that Mary is to the right of Tom arises. We obviously *perceive* this new fact when these two pieces of information are conjoined. In reality, when some facts are accumulated, a new fact comes out of the accumulation of the facts. There is nothing amalgamating the facts. A photograph of reality reveals the result of conjunctive facts without any connection symbol. Let us see how more than one piece of information is combined in diagrammatic and linguistic systems.

Suppose I have the following representation that represents the information that Susan is to the right of Tom: (Let "□" denote Tom, "○" Susan, and "△" Mary.)

Suppose I obtain another piece of information, Mary is to the right of Susan, as follows:

All I will do is add this new piece of information. First, I do not need to bring in any extra device to add this new information. Second, when the second piece of information is added to the first one, a new piece of information, that Mary is to the right of Tom, is perceived without any manipulation of the representation. Let us recall two kinds of diagrams used in cases 2 and 4. Case 2 represents individuals by geometrical figures and case 4 by the letters of the alphabet. However, with respect to the way conjunctive information is displayed, the representations of case 2 and case 4 are not different from each other at all. This is very similar to the way more than one piece of information is displayed in reality as described before.

Suppose I want to combine the following two pieces of information given in a linguistic text:

$$RightOf(s, t),$$
$$RightOf(m, s),$$

where "t" denotes Tom, "s" Susan, and "m" Mary.

In case 3 of our example (i.e., the first-order sentence), these two pieces of information are connected with the symbol $\wedge$, and we get the following sentence: $RightOf(s,t) \wedge RightOf(m,s)$. Compared with diagrammatic systems, at least two different things are taking place here. First, a new syntactic device, $\wedge$, was introduced. Second, the resulting formula does not reveal the new information, that Tom is to the right of Mary, in a perceptually obvious way. In order to extract this information, we need to know how to manipulate this resulting sentence, $RightOf(s,t) \wedge RightOf(m,s)$.

The conjunction symbol $\wedge$ is such a pure convention that we cannot find anything in reality that corresponds with it. As we said earlier, more than one piece of information is displayed in reality without any mediating object. Accordingly, how a symbol looks is absolutely arbitrary. It could be "&," "·," "⋆," or "⊗." Some first-order languages allow two sentences to be connected without any syntactic device, for example, pq, where p and q are sentences. However, even in this case, I do not think that this language shows a diagrammatic feature. Even though two sentences are combined without any syntactic device, the resulting sentence does not reveal new information in a perceptually obvious way. For example, $RightOf(s,t)RightOf(m,s)$ would be a resulting sentence that combines the previous two pieces of information. This sentence is not different from sentence $RightOf(s,t) \wedge RightOf(m,s)$, except that they adopt different conventions. The former sentence can be said to adopt an invisible syntactic device, whereas the latter adopts a visible one, $\wedge$.

This conforms to our earlier statement that how this conjunctive symbol looks is arbitrary.

This discussion shows that an important difference between the way conjunctive information is represented by diagrammatic and linguistic systems is not only whether or not a new syntactic device is introduced, but whether or not the resulting diagram or sentence reveals new information in a perceptually obvious way.

These clear differences in the way conjunctive information is handled explain our intuitive distinctions between the two kinds of systems. We have already seen that maps, timetables, or family trees are diagrammatic in that relations among objects are represented in terms of spatial arrangements despite the fact that some of the represented relations are not spatial. These examples also show the characteristic of conjunctive information being represented. When we put more information in a timetable, we do not introduce any syntactic device but rather supplement the table by putting in a new piece of information. We do not need any additional convention for adding new pieces of information, unlike with linguistic systems. A similar explanation works for family trees or maps. We also have an intuition that pie charts are diagrams, not linguistic texts, despite the fact that letters and figures are written to convey crucial information. We add information in a pie chart in the same way as we do in other diagrams.

The capacity for expressing conjunctive information in this way is one of the strong points of successful diagrammatic representation systems, compared with symbolic systems. For example, transitive, symmetric, or asymmetric relations are represented in a perceptually obvious way when more than one piece of information is added. We know this convenience very well. This is one of the main reasons why we use pictures or diagrams as a heuristic tool and why we transform a linguistic text to a diagram, for example, when we solve problems on the Graduate Record Exam. For example, in the case of maps or family trees, many transitive relations, for example, "being north of," "being left of," "being ancester of" or "being descendant of," are perceptually inferred without any manipulation. In linguistic texts, more than one piece of information is enumerated in a linear mode or mediated by a conjunction symbol. In the process of translation to visual representation, the linear mode is changed into a spatial relation and all conjunction symbols are dropped.

6.2.3 Tautologies and contradictions

Tautologies and contradictions are not the most interesting cases in a representation system. Nonetheless, we expect a reasonable representation system to represent tautological and contradictory information. These two kinds of information are quite different from other contingent information. First, does a tautological sentence or a contradictory sentence convey any information? Suppose some one said, "It is raining or it is not raining." Did his statement contain any information? Suppose someone said, "It is raining and it is not raining." Did he convey any information? Many of us would think that in the former case there is no information conveyed and in the latter his information is always wrong. Standard model theory captures this intuition in the following way: A sentence is logically true if and only if every model satisfies it,[4] and a sentence is a contradiction if and only if no model satisfies it. As we know, in many cases, to tell whether a sentence is a tautology or a contradiction is not trivial. It requires more than simple perceptual inferences. I claim that diagrammatic systems represent tautologies and contradictions in a more perceptually obvious way than linguistic systems do.

Wittgenstein, in his *Tractatus*, classifies tautologies and contradictions as non-standard propositions. Both tautologies and contradictions are treated as degenerate cases of propositions which do not fit in his picture theory of language. If propositions are pictures of reality that reflect the logical forms of reality, of what kind of reality are tautologies or contradictions pictures? Wittgenstein's answer is that there is no reality of which either a tautology or a contradiction is a picture. Tautologies admit every possible situation and contradictions do not admit any possible situation. Since there is no reality that admits all possible situations or no situation, neither a tautology nor a contradiction can be a picture of reality. That is, "They [tautologies and contradictions] do not represent any possible situation" [28], 4.462. Similarly, Wittgenstein makes the same comments about both tautologies and contradictions in several other places, especially when he says, "tautologies and contradictions show that they say nothing" [28], 4.461, or, "Tautologies and contradictions lack sense" [28], 4.461.

I agree with Wittgenstein that both tautologies and contradictions may together be called nonstandard propositions. However, there is a difference between these two atypical propositions. Tautologies convey no information, whereas contradictions convey multiple pieces of infor-

[4] We assume that every tautology is locally true.

mation that conflict each other. Accordingly, tautological information is vacuously true, whereas contradictory information is always false. I claim that these aspects are conveyed in diagrams in a more perceptually obvious way than in linguistic texts.

In the case of diagrams, if a diagram contains no representing fact, then it is a tautological diagram. We can draw a diagram without any representing fact. A pie chart or a timetable or a map without any content in it is a tautological diagram. It does not convey any information. However, we cannot find an analogous case to this in a linguistic system, since a tautological sentence is still a nonempty string. We can write down a long string of symbols that does not convey any information. The formula $(A \rightarrow (B \rightarrow C)) \leftrightarrow ((A \wedge B) \rightarrow C)$ is an example.

That contradictions are represented more perceptually in diagrams than in linguistic texts is closely related to the way conjunctive information is represented in diagrams, as discussed in the previous subsection. More than one piece of information is added without any syntactic device in diagrams, whereas they are connected with conjunctive symbols in linguistic texts. In the latter case, whether an added new piece of information is contradictory to the existing one or not does not make any difference. However, in the case of diagrams, this conflict becomes immediately noticeable, since conflicting spatial arrangements are displayed in an obvious way.

Suppose the following three pieces of information are given:

(1) Susan is to the left of Tom.
(2) Tom is to the left of Mary.
(3) Mary is to the left of Susan.

A first-order language will give us the following sentence:

$$LeftOf(s,t) \wedge LeftOf(t,m) \wedge LeftOf(m,s).$$

This is a contradiction and we can prove it by using the fact that the relation "being left of" is transitive and asymmetric. However, this contradiction is not obviously revealed in this conjunctive sentence. This is why we need to manipulate this sentence to show the contradiction.

If we display these three pieces of information in a diagram, the following is what we get: Let "○" denote Susan, "□" Tom, and "△" Mary.

○ □ △ ○

Or,

Susan Tom Mary Susan

Both diagrams clearly violate the following spatial constraint of this representation: No geometrical figure or name occupies more than one place simultaneously. The contradiction is perceptually obvious.

Timetables are a good example for using this diagrammatic feature in a crucial way. A timetable is a diagrammatic representation of a schedule. One of the main reasons why we want to represent a schedule is to avoid any conflict among appointments. Suppose we want to add a new appointment to a timetable. If this new appointment conflicts with any existing appointments, this conflict will be revealed clearly, since this addition would violate the following spatial constraint of this representation system: No space can be occupied by more than one appointment. Any conflict, if there is one, is perceptually obviously represented. This is one of the reasons why this representation is more used than linguistic representation for arranging a schedule.

Arrangements of a dining table or of offices are represented quite often in terms of diagrams. A conflict, that is, sitting more than one person in the same chair or assigning to an office more than the number of people that the office can accommodate, should be a crucial problem. As seen so far, a diagram helps us to avoid this kind of a conflict easily, by revealing contradictions in a more perceptually obvious way.

6.3 Venn-I as a diagrammatic system

In this section, by examining how the three diagrammatic features discussed in the previous section are manifested in the Venn-I system, I will support my assumption that Venn-I is a diagrammatic system.

6.3.1 Relations of sets

In the first section of this chapter, I claimed that the nature of the primitive objects of a system is not intrinsically different between diagrammatic and linguistic representations. I showed that whether geometrical figures or letters are used cannot be a good criterion for making a distinction between these two kinds of systems. Both in the Euler and Venn systems, we sometimes annotate circles with letters of the alphabet. I also discussed a case in which a system aims to represent abstract entities. In this case, it is impossible to come up with primitive objects that look like these entities. No one would expect Euler and Venn circles to look like what these circles represent, that is, sets.

In this subsection, I explain a difference between Venn-I (and the Euler system) and a first-order set-theoretic language in the way two kinds of relations, that is, the membership relation and the relations among sets, are represented in each system.

In Venn-I, we chose a closed curve and a rectangle as the primitive objects to represent sets. In a first-order set-theoretic language, we chose letters of the alphabet (for example, $v_1, v_2, \ldots$ as variables and $A_1, A_2, \ldots$ as constants) to represent sets. Neither the geometrical figures nor the letters can look similar to the sets each system aims to represent. In this sense, both systems adopt their own conventions: In Venn-I we adopt a convention that assigns a set to a basic region.[5] In the case of a first-order language, we adopt the convention that a set is assigned to each variable or each constant of the language. Up to this point, both diagrammatic and linguistic systems seem to rely on a pure convention which is arbitrary. Accordingly, there seems to be no necessary relation between a set and a closed curve any more than there is between a set and a variable symbol, whereas a necessary relation holds between Tom and the picture of Tom. We cannot perceptually infer the representation of a set from either a closed curve or a letter of the alphabet, whereas we can infer the representation of Tom (i.e., infer how he looks) from his picture.

However, there seem to be more constraints on the choice of primitive objects in diagrammatic systems than in symbolic systems. Depending upon the conventions of a system, some constraints follow. For example, as mentioned earlier, Venn-I adopts the following convention: A set is assigned to a region enclosed by some geometrical figure (in this case, a closed curve or a rectangle). This convention requires that this geometrical figure should be a closed one so that it may create an interior region and an exterior region. For example, the following figures do not enclose an area:

[5]According to this convention, identical-looking closed curves might represent different sets.

Accordingly, they cannot make a distinction between the area enclosed by this figure and the area that is not.[6] In the case of linguistic systems, sets are assigned to symbols. It really does not matter what those symbols look like.

Neither Venn-I nor the Euler system has a syntactic object that denotes an individual, that is, a member of a set. Suppose we use dots for this purpose. In this case, we introduce new syntactic objects which are as arbitrary as letters of the alphabet in a first-order language. With these two syntactic devices, that is, dots and circles, we want to represent a membership relation between an individual and a set.

Suppose the region enclosed by closed curve A (let us name it region A) represents the set of animals:

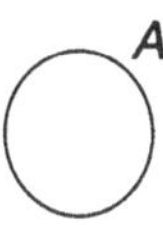

We want to represent an individual that is an animal. Where a dot (which represents this individual) should be drawn seems to be quite obvious. Since we adopt the convention that a region represents a set, this dot should be drawn inside circle A. If the dot is inscribed outside the circle, we seem to violate the convention about the representation of sets. A set is assigned to the region enclosed by a circle, not to the region excluded by a circle. Accordingly, we want to say that every animal should correspond to a location inside the closed curve and that every nonanimal is assigned a location outside the closed curve. These assumptions can be captured in the following convention:

> Every object x in the domain is assigned its unique location, say, l_x, in the plane such that l_x is *in* region R if and only if x is a member of the set region R represents.

This convention is not absolutely arbitrary because it displays the following concept of membership: If object a is a member of set A, then we say a is *in* set A. However, this is still a convention in the sense that this

[6] Of course, we could adopt a different convention. For example, a set is assigned to a different geometrical figure. Lambert's linear diagrams (introduced in Chapter 2) seem to adopt the convention that a set is assigned to each line (not to the area a line encloses).

is not the only way to visualize the concept of "being a member of." For example, Lambert's system adopts a different mode of visualization. He visualizes individuals as points and imagines all the individuals in a row. A concept is assumed to be represented as a line covering the individuals that belong to the concept. Accordingly, the collection of animals, for example, can be represented as the line that is drawn over those points each of which represents an animal. Quite naturally, he chose a line to represent a collection, since a line is a collection of points just as a set is a collection of individuals. We can express Lambert's convention in a way very similar to that we did for the Euler and the Venn systems:

Every object x in the domain is assigned its unique location, say, l_x, in the plane such that l_x is *on* line L if and only if x is a member of the set line L represents.

One of the characteristics of diagrammatic systems discussed in the previous section is that diagrams represent relations (whether they are spatial or nonspatial) among objects in terms of spatial relations. The membership relation represented in the Euler, the Venn, or the Lambert system is a good example for this characteristic. This relation is not spatial, but these diagrammatic systems represent it in terms of a spatial (i.e., either *in* or *on*) relation. Either spatial relation, that is, being in a circle or being on a line, appeals to our natural perceptual ability. Accordingly, both ways of capturing the concept of membership are less arbitrary than the way a first-order language of set theory does. In the latter case, we need one more syntactic device for the membership relation. For example, $a \in A$ if the object denoted by 'a' is a member of the set denoted by "A." Adopting this symbol, "$\in$" is a pure convention. It could be a symbol of other shape, for example "$\rightarrow$," "$\leadsto$," "$\odot$."

A difference in degrees of arbitrariness between linguistic and diagrammatic systems becomes more obvious when we try to represent relations of sets, that is, the intersection, the union, and the differences of sets, which are also sets. First I will show that Euler does not need an extra syntactic device for this representation, unlike with linguistic systems. Let us return to some Euler circles:

Suppose that we accept two of the conventions discussed already: (i) A set is assigned to a region enclosed by a closed curve. (ii) Every object x

in the domain is assigned its unique location, say, l_x, in the plane such that l_x is in region R if and only if x is a member of the set region R represents. And suppose that the smallest circle within which "a" was written denotes a set for being a and the circle with "b" a set for being b. Then, it follows that the diagram on the left represents the fact that all a is b and that the diagram on the right side the fact that no a is b.

Suppose Euler made the opposite suggestion as follows: The left-hand diagram represents "No a is b" and the right one "All a is b." This would violate at least one of those two conventions. That is, how a relation between two sets is represented is *not* a new convention to be introduced, *but* a consequence of the conventions about syntactic objects that are already established.

How does the Venn system represent the relations among sets? As seen in the second chapter, Venn introduces a primary diagram as follows: We draw closed curves in a certain way so that they partially overlap each other only once. According to this convention, two sets, say, the set of unicorns and the set of red things, are represented as follows:

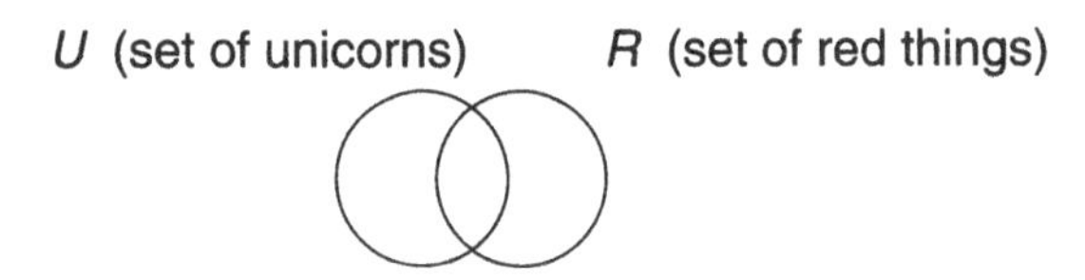

After accepting the conventions mentioned so far, referring to some relation of these two sets, for example, the intersection of the set of unicorns and the set of red things, is not arbitrary any more. Region *U-and-R* (the region that is the intersection of region U and region R) *should* represent the intersection of these two sets, *given* that U represents the set of unicorns and R the set of red things. According to the convention that the unique location of each object is determined by the membership of the given sets, the intersection of region U and R is occupied by all and only objects that belong to both sets. Otherwise, we would find this system contradicting its own existing conventions. Suppose that the region $U - R$ represented the intersection of the set of unicorns and the set of red things. Suppose an object a is represented in this region. Then, by the second assumption [7] and by our previous assumption, a is a member of both the set of unicorns and the set of red things. But, a is not in the region enclosed by closed curve R. Again, by the second

[7] Every object x in the domain is assigned its unique location, say, l_x, in the plane such that l_x is in region R if and only if x is a member of the set region R represents.

assumption, a is not a member of the set of red things. This contradicts what we just said: a is a member of both the set of unicorns and the set of red things. It is very important to realize that our extension $\bar{s}$, of set assignment s (which was formalized in §3.3.1) is not conventional, but a logical consequence of the conventions about the representation of sets.

In the case of a linguistic representation system, we need new symbols for the relations of these two sets, just as we introduced a new symbol for membership, $\in$. For example, $A \cap B$ represents the intersection of set A and B, $A \cup B$ the union of them, and $A - B$ the difference between them. Choosing "$\cap$" for the intersection, rather than "$\cup$" or "$\leftrightarrow$," is absolutely arbitrary. Even after we learn the meaning of "$\in$," we have to learn the meanings of "$\cap$," "$\cup$," and "$-$" to represent the relations of sets. These symbols are as arbitrary as symbols for individuals or for sets. The only difference is that symbols for individuals and for sets are constants, whereas symbols for membership and for set relations are predicates. We need to learn the semantics for these predicates just as we do for the constants.

However, in either the Euler or the Venn system, we do not need any other convention to represent membership or set relations, after accepting the conventions on the representation of sets and individuals. All of these relations are represented in terms of spatial relations of representing objects. As just seen, these spatial relations are natural results of the conventions about the existing symbols. Accordingly, we can infer perceptually (without much intellectual effort) what these diagrammatic systems aim to represent. That is, there is a more natural correspondence between the definition of $\bar{s}$ (given set assignment s) and the relations of sets, than between the definitions of $\in$, $\cup, \cap$, or $-$ and the set relations. This correspondence, which shows a clear difference in degrees of perceptual inferences between diagrammatic and linguistic systems, is one piece of evidence to support my claim that Venn-I is diagrammatic.

6.3.2 Conjunctive information

In this subsection, I will provide one more piece of evidence for Venn-I being a nonlinguistic system, by showing that the way Venn-I conveys conjunctive information is different from the way a linguistic representation does.

In Chapter 2, we saw how Venn improved Euler diagrams to let partial knowledge be illustrated in a diagram. When our knowledge about the relations of the represented classes is increased, we just put in more marks, that is, shading or x's, in the appropriate compartments. One of the best results from this improvement is that we can accumulate information in one diagram. In other words, we can express conjunctive information much more freely than in the Euler system.

Let us compare the first-order language $\mathscr{L}_0$ of Chapter 5 and Venn-I. Suppose pieces of information I_1 and I_2, each of which is representable both in $\mathscr{L}_0$ and in Venn-I, are given. Let φ_1 and φ_2 be the well-formed formulas of $\mathscr{L}_0$ that respectively represent I_1 and I_2. Let D_1 and D_2 be the well-formed diagrams of Venn-I for I_1 and I_2, respectively. Suppose that we want to combine these two pieces of information to formulate a meaningful unit of each system.

In the case of $\mathscr{L}_0$, we connect φ_1 and φ_2 by a conjunction symbol $\wedge$. Hence, "$(\varphi_1 \wedge \varphi_2)$" is a meaningful unit of $\mathscr{L}_0$ which represents the conjunction of I_1 and I_2. What takes place here is that two *wffs* are juxtaposed in a linear mode mediated by a new syntactic object, $\wedge$. This new arrangement does not reveal new information any more than the two separate *wffs*, φ_1 and φ_2, do.

In the case of Venn-I, we can think of two different cases for conjoining these two pieces of information.

(i) We unify D_1 and D_2 by the rule of unification to get a new diagram, say, D'.

(ii) As we pointed out in the second chapter, the Venn system allows us to add new information to a given diagram. Accordingly, we may add a closed curve, a shading, or an x-sequence to D_1, depending upon I_2, to get a new diagram, say, D''.

The two processes seem to be different from each other. In the first case, we draw a separate diagram for I_2 and unify it with the existing diagram D_1. In the second, we do not draw a separate diagram for I_2 but add this new piece of information to the existing diagram D_1. However, we know that D' and D'' are equivalent to each other.

There are two points we should notice. First, in either case (i.e., (i) or (ii)) Venn-I did not introduce any syntactic device for combining two pieces of information, whereas $\wedge$ is used in a first-order language. Second, the resulting diagram, either D' or D'', has a new representing fact which neither D_1 nor D_2 has, whereas $(\varphi_1 \wedge \varphi_2)$ is just a linear arrangement of the two formulas. Let me illustrate these important differences through the following example.

Example 59 *Let* I_1 = "Every unicorn is red," *and* I_2 = "No unicorn is red." *From these two pieces of information, we want to extract the information, say, I, that there is no unicorn.*

In the case of $\mathscr{L}_0$, $\varphi_1 = \forall x(Ux \rightarrow Rx)$ and $\varphi_2 = \forall x(Ux \rightarrow \neg Rx)$. What we want to get is formula "$\neg\exists xUx$." The conjunctive information of I_1 and I_2 is represented as "$(\varphi_1 \wedge \varphi_2)$," that is, "$(\forall x(Ux \rightarrow Rx) \wedge \forall x(Ux \rightarrow \neg Rx))$." However, this conjunctive formula does not contain formula "$\neg\exists xUx$" as one of its parts. This conjunctive sentence does not contain the conclusion in a perceptually obvious way. Therefore, we need to manipulate $(\varphi_1 \wedge \varphi_2)$ by rules of inference.

The conjunction symbol of $\mathscr{L}_0$, $\wedge$, is just an artificial device to represent conjunctive information. In order to see the arbitrariness of this device, we can always replace it with a symbol of other shape, for example, "·," "&." Such a device does not change the structures of the given formulas but rather connects them in a linear mode. Hence, this connected formula does not reveal new information any more than the given two original formulas do. If the information we want to extract is not obviously revealed in the original formulas, the formula with a conjunction symbol between the given formulas does not reveal the information in an obvious way either. In the previous example, $\neg\exists xUx$ is not a component of φ_1 or of φ_2. Accordingly, $\neg\exists xUx$ is not a component of $(\varphi_1 \wedge \varphi_2)$ either.

On the other hand, when we combine pieces of information in Venn-I, a new representing fact arises out of the combination of the given pieces of information without any special device added. The two following cases can be considered:

(i) D_1 and D_2 represent I_1 and I_2, respectively. We unify these two into D'.

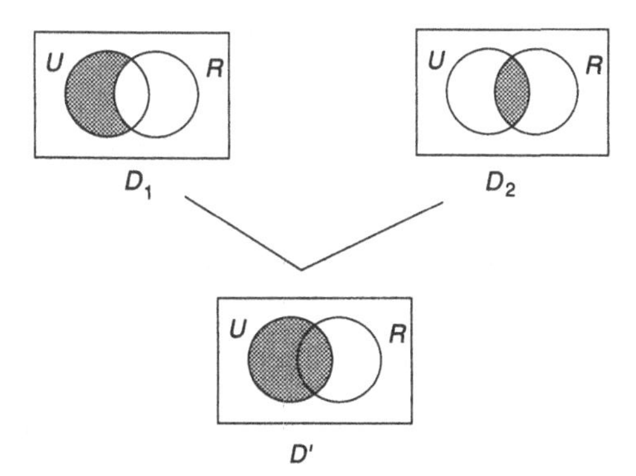

(ii) D_1 represents information I_1. Add information I_2 by shading region *U-and-R* to get D''.

What we want to get is a diagram D such that $\langle\langle \text{Shading}, U; 1\rangle\rangle \in RF(D)$. Both D' and D'' have representing fact $\langle\langle \text{Shading}, U; 1\rangle\rangle$. An interesting point is that neither D_1 nor D_2 has $\langle\langle \text{Shading}, U; 1\rangle\rangle$ as a representing fact. Unifying the given pieces of information, or accumulating the given pieces of information, we *see* the representing fact we want to know in a resulting diagram. This convenient way of combining the given pieces of information makes it possible for us to *read off* the conclusion of an argument if the argument is valid. There is not much need for manipulation, unlike with linguistic representation. As said in the previous section, this is one of the main reasons why diagrams are used in our reasoning.

6.3.3 Tautologies and contradictions

In this subsection, I will show how Venn-I represents tautologies and contradictions, and will claim that these two special cases of information are represented in a more perceptually obvious way in this system than in linguistic texts.

Let a tautological diagram of Venn-I be a diagram supported by every situation. Let Ω be the set of tautological diagrams. Then, Ω is the smallest set of *wfd*s satisfying the following:

(1) Any rectangle drawn in the plane is in set Ω.
(2) If D is in set Ω, then if D' results from adding a closed curve interior to the rectangle of D by the partial-overlapping rule, then D' is in set Ω.

In Venn-I, there are only two kinds of representing facts. They are whether a region is shaded and whether a region has an x-sequence. Each member of set Ω has no shading or x-sequence in any of its regions. Therefore, every tautological diagram has no representing fact. Wittgenstein points out that a tautology admits every possible situation. The semantics of Venn-I shows that a diagram is tautological if and only if every possible set assignment satisfies it. That is, $\{D \mid RF(D) = \emptyset\} = \{D \mid \forall s(s \models D)\}$. The fact that tautologies do not give us any information, that is, do not say anything, is perceptually obvious in this diagrammatic system.

A diagram of Venn-I is a contradictory diagram if it has a region with both a shading and an x-sequence. Any contradictory diagram has at least two representing facts, such as $\langle\langle \text{Shading}, A; 1\rangle\rangle$ and $\langle\langle \otimes^n, A; 1\rangle\rangle$ for some region A. If a contradictory diagram were satisfied by a set assignment, say, s, then s would assign to the same region both the empty set and a nonempty set. Therefore, there is no set assignment that satisfies a contradictory diagram.

The Venn-I system represents tautologies and contradictions by manifesting the following properties of these concepts: If what someone says is a tautology, we know that what he says is true but not informative at all. We treat his statement as if he did not say anything despite the fact that he uttered something. Tautologies are more like silence from which we cannot extract any information. They represent no information. For example, suppose I was commanded "to read this newspaper or not to read it." This is the same as if I did not get any command, since whatever I may do I carry out the command. This intuition is reflected very well in tautological diagrams of Venn-I, since they do not carry any representing facts.

If what someone says is a contradiction, we know that there is no situation in which what he says is true. Suppose I was told "to read the newspaper and not to read it." There will be no situation in which this command will be carried out. Contradictions represent conflicting information. Unlike tautologies, they seem to represent something, but it is impossible for whatever they represent to be true. A diagram of Venn-I that has a region with both a shading and an x-sequence reflects this intuition. It is impossible for the same region to be assigned the empty set and a nonempty set by the same set assignment s, since s is a function.

To sum up: In the case of the concept of "tautology," the Venn-I system visualizes one characteristic of this concept, that is, the carrying of no information, in terms of a diagram with no representing fact. In the case of the concept "contradiction" the incompatibility or the inconsistency (as a property of this concept) is visualized in that both a shading and an x-sequence, which carry opposite meanings, occupy the same region of a diagram.

6.4 Linguistic elements in Venn systems

We have seen that relations of objects and more than one piece of information are represented in diagrams without any new syntactic device. We have also discussed that this helps a system reveal information in a perceptually more obvious way. However, some skeptic might turn this strong point of diagrammatic representation systems into a defect of these systems. He might raise the following questions: Is a visual system able to represent the negation of given information? How about disjunctive information? How about hypothetical information? At least in the diagrammatic method Venn presented, the answers to these questions are negative. As seen in the second chapter, Venn's original diagrammatic system is able to represent only universal propositions. Since the negation of universal propositions are existential propositions, negative information cannot be represented in his system. Also, even though his system can represent "All *A* is B" and "All *B* is *A*," respectively, it cannot represent "All *A* is B or all *B* is *A*." Accordingly, we cannot represent the following hypothetical information in this system either: that if all *A* is *B*, then all *B* is *A*.

Venn himself discusses these matters in the last chapter of his book, *Symbolic Logic*. At the end of the discussion, he almost seems to conclude that diagrams are not equipped to handle hypothetical or disjunctive information. Let me quote two passages which are his final remarks for each case:

> Those who adopt the judgment interpretation can hardly in consistency come to any other conclusion than that hypotheticals are distinct from categoricals, and do not admit of diagrammatic representation.[8]

> As will be gathered from the above, the attempts to give diagrammatic illustration of the disjunctive relation have been extremely few, and these few are mostly under arbitrary conditions.[9]

Venn did not try to explain why these kinds of information do not admit a diagrammatic representation. The first reason might be the lack of a syntactic device to represent existential propositions. Let us introduce one of Peirce's new syntactic devices, "x," for existential statements. Then, in a special case, we can represent the negation of given information. For example, given the figure on the left, we can draw the right one to represent the negation of the information that the left one conveys:

[8]Venn [27], p. 521.
[9]Venn [27], p. 524.

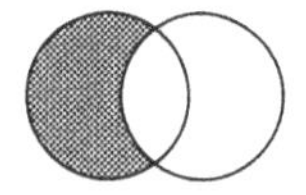

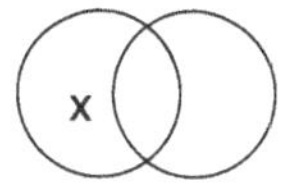

However, if we want to negate the information conveyed by the following diagram, we would need to represent disjunctive information.

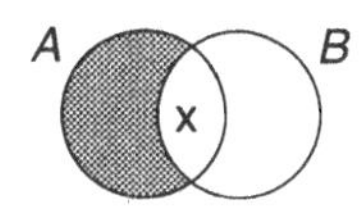

Even with "x," Venn's original method does not have a way to represent the information that some A is not B or no A is B.

The lack of syntactic devices in visual systems, our skeptic might claim, is responsible for this defect. In the case of linguistic representation systems, the syntactic devices of each system ("¬," "∨," "→," or "it is not the case," "or" "if ... then") can handle these kinds of information. As seen in the second chapter, Peirce, who was aware of this limit of Venn diagrams, introduced a syntactic device to overcome this defect. The device is a line, which connects marks "o" and "x."

I think these questions raised by our skeptic should be taken seriously. As he thinks, they might reveal a crucial drawback of diagrams as representation systems. Or, these questions might lead us to a fundamental difference between linguistic and nonlinguistic representation systems and help us decide which system should be adopted for given information. Or, these questions about different representation systems might turn into a question about different kinds of information, that is, conjunctive versus disjunctive information. I think all these issues are closely related to each other.

In the second section of this chapter, I claimed that diagrammatic systems are more similar to reality than linguistic systems in the way conjunctive information is displayed. I also argued that this similarity is characterized by the need for fewer conventions and the power of perceptual inferences in diagrams. In the following, I will show that disjunctive information is intrinsically different from conjunctive information and that disjunctive information is not displayable in reality. This observation leads us to the following prediction: Any system that seeks to represent disjunctive information needs to bring in an artificial syntactic device with its own convention. Accordingly, the system will not carry the power of perceptual inferences for this kind of information.

6.4.1 Disjunctive information

Consider the following disjunctive information: Mary is to the left of Tom or is to the right of Tom. Suppose we take a picture to verify this information. If this information is correct, we will get *either* a picture of the situation in which Mary is to the left of Tom *or* a picture of the situation in which Mary is to the right of Tom. We will never get a picture of the situation in which the disjunctive fact that Mary is to the left of or to the right of Tom is displayed. Interestingly, either of the pictures we might get conveys only one disjunct of the original disjunctive information. That is, the picture will carry more than the original disjunctive information. A situation displays only conjunctive information. We can find a situation that supports disjunctive information. But this situation contains more than the disjunctive information.

This is why Venn-I needs to introduce a new syntactic device, "–," to represent disjunctive information, unlike with the representation of conjunctive information. However, "–" of Venn-I is different from the disjunction symbol $\vee$, of $\mathscr{L}_0$ in the following sense: While $\mathscr{L}_0$ may mediate any two pieces of information representable in this system by $\vee$, Venn-I can represent the disjunction of some (not all) pieces of information. The following pieces of information are representable both in Venn-I and in $\mathscr{L}_0$:

I_1 = "Some unicorns are red."

I_2 = "Some visible things are red."

I_3 = "All unicorns are red."

$\mathscr{L}_0$ can represent the disjunction of any combination of these pieces of information. On the other hand, the disjunction of I_1 and I_2 is representable in Venn-I, but not the disjunction of I_1 and I_3 or the disjunction of I_2 and I_3.

In Chapter 4, Venn-I was extended to represent the disjunction of any pieces of information representable in Venn-I. In this extended system, Venn-II, the line of Venn-I, that is, "–," is used in a way very similar to the way $\vee$ is used in $\mathscr{L}_0$. That is, "–" connects *wfd*s of Venn-II and, accordingly, Venn-II may mediate any two pieces of information representable in Venn-II by this syntactic device.

The disjunction sign "–," gives both Venn-I and Venn-II a power to represent disjunctive information. It should be noticed that the introduction of this device is purely conventional, as in the case of $\wedge$ and $\vee$ of symbolic languages. Accordingly, both Venn-I and Venn-II lose their visual power after using this arbitrary device. This point is illustrated very well in terms of two kinds of uses of the verb *show* in the next subsection.

6.4.2 Primary versus secondary "show"

In *Situations and Attitudes*, Barwise and Perry make a distinction between two kinds of uses of *see*, following Dretske's distinction in his *Seeing and Knowing*. The primary use of *see* is for reporting what we see through perception (they express it as see_p), and the secondary use is for reporting what we know based upon what we see in the primary sense (they say see_s). They demonstrated these two different uses of *see* with the following sentence:

I saw_p that the tree was whipping around, so I saw_s that the the wind was blowing.[10]

One observation they make about the verb *see*, while carrying out their aim to develop the semantics of these two uses of *see*, is that the following disjunction distribution principle does not hold:

(DDP) If a sees that $P \vee Q$, then a sees that P or a sees that Q.

They suggested the following sentence as a counterexample to this principle:

(a) Sarah *saw* that Jim or Joe ate her egg (b) but she didn't *see* that Jim ate it or that Joe ate it.[11]

In order to get a natural reading of this sentence, *saw* of (a) should be saw_s and *see* of (b) should be see_p. That is, we can see_s some disjunctive information without $seeing_p$ either of the disjuncts. I use Barwise and Perry's observation on the verb *see* to make the same kind of distinction for the verb *show*. That is, the primary use of *show* (call it $show_p$) is for reporting what is shown to our perception, and the secondary use (call it $show_s$) is for reporting what is known based upon what is shown in the primary sense. This difference is demonstrated in the following sentence:

Sarah never $shows_p$ her tears to any one. This $shows_s$ how strong she is.

Let us start with the case of pictures to see how the principle of disjunction distribution (DDP) works for the verb *show*. We will test whether or not the following principle holds for a picture:

[10] Barwise and Perry [5], p. 194
[11] Barwise and Perry [5], p. 205

(DDP) If a picture shows that $P \vee Q$, then it shows that P or it shows that Q.

Suppose we took a picture of a classroom. Since no disjunctive information is displayable in reality, this picture cannot show$_p$ any disjunctive information. In reality, disjunctive information can be only inferred (not directly displayed) either from some inferred information or from some displayed information. Therefore:

If this picture shows that $P \vee Q$, then the picture shows that P or the picture shows that Q.

This observation corresponds to one of the points in the last subsection that given a piece of disjunctive information a picture always carries more than the original disjunctive information.

Let us see whether DDP for the verb *show* is true of the Venn systems, depending on whether or not the disjunction sign "–" is introduced into the systems. The observation that will be made in the following paragraphs supports my claim that the introduction of "–" adds a new property to the Venn systems, which makes these systems essentially different from photographic representation.

What a diagram shows is either that some set is empty or that some set is not empty. Before the disjunction sign "–" is introduced into the Venn systems, DDP is true. And after we use "–" in connecting ⊗'s, this principle works only for some propositions. However, after "–" is used in connecting diagrams, this principle does not hold any more.

Suppose D is a diagram without a disjunction syntactic device "–." That is, D is a diagram of Venn-I that does not have an x-sequence with more than one ⊗. The situation is very similar to the photographic representation we discussed earlier. Since D can no more represent disjunctive information no more than a photograph can display it, any disjunctive information cannot be *shown*$_p$ in D, but can be *shown*$_s$ in D. Accordingly, D cannot *show*$_s$ a piece of disjunctive information without *showing*$_p$ one of the disjuncts of this information. For example, what the following diagram *shows*$_p$ is that the set region $A - B$ represents (let us say $\bar{s}(A - B)$) is not empty:

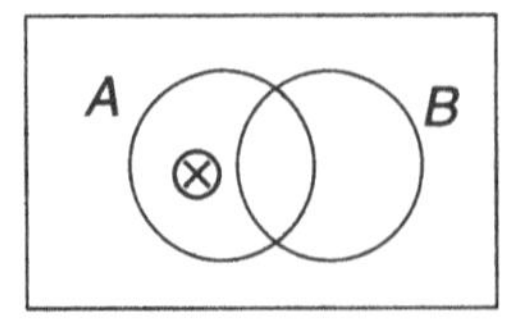

Accordingly, it also shows$_s$ the following: $\overline{s}(A - B)$ is not empty or $\overline{s}(A\text{-}and\text{-}B)$ is not empty. Therefore, just like with photographic representation, the following principle works for diagram D without "–."

If diagram D shows that $P \vee Q$, then D shows that P or D shows that Q.

Suppose D' is a diagram of Venn-I that has an x-sequence with more than one ⊗. Now, some disjunctive information, say, $P \vee Q$, can be represented without P or Q being represented. For example, the following diagram shows that $\overline{s}(A - B)$ is not empty or $\overline{s}(A\text{-}and\text{-}B)$ is not empty.

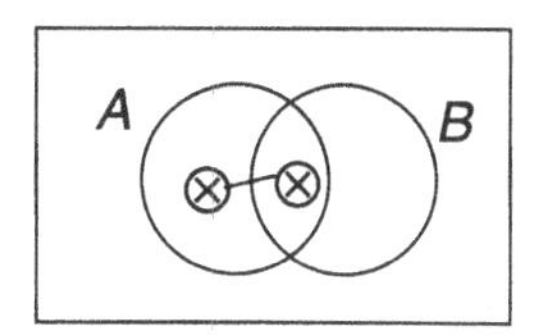

However, this diagram shows neither that $\overline{s}(A - B)$ is not empty nor that $\overline{s}(A\text{-}and\text{-}B)$ is not empty, so DDP does not hold here. Venn-I, however, has limited power to represent disjunction. Specifically, it cannot handle the disjunction of facts about the emptiness of sets. Therefore, the following principle still works for information about the emptiness of sets. That is,

let D' be a diagram of Venn-I and P and Q each be the proposition that some set is empty. Then, if diagram D' shows that $P \vee Q$, then D' shows that P or D' shows that Q.

When we move to Venn-II, we can predict that this disjunction distribution principle holds neither for information about the emptiness of sets nor for information about the nonemptiness of sets. The following is an example that shows that disjunctive information is representable without each disjunct being represented:

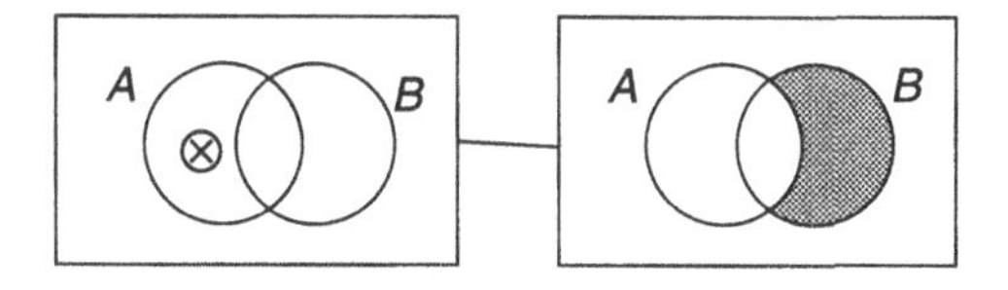

This diagram shows that $\overline{s}(A - B)$ is not empty or $\overline{s}(B - A)$ is empty, but does not show either of the disjuncts. DDP does not hold in Venn-II for either type of information, that is, information about the emptiness of sets or about the nonemptiness of sets. The situation is quite similar to a linguistic text. For example, formula "$\forall x(Px \rightarrow Qx) \vee \exists x(Px \wedge Qx)$"

shows that every P is Q or some P is Q, but does not show either of the disjuncts. This demonstrates clearly that the introduction of a syntactic device, "–," makes the Venn systems less perceptually obvious than mere photographic representation.

7
Conclusion

In the introduction, we identified a general prejudice against diagrams in the history of logic and mathematics. Diagrams, in spite of their widespread use, have never been permitted as valid or real proofs. We also identified as one of the main reasons behind this prejudice a general worry that diagrams tend to mislead us. I showed in the main part of this work that the misapplication of diagrams is not intrinsic to the nature of diagrams. Venn diagrams, one of the most well-understood and widely used kinds of diagrams, can be presented as a standard representation system which is sound and complete. Accordingly, as long as we follow the transformation rules of the system, the use of Venn diagrams should be considered a valid or real proof, just as the use of first-order logic is. So, mathematicians' and logicians' worries about the misapplication of diagrams in general cannot be justified. We should not give up using diagrams in a valid proof just because there is a possibility of the misuse of diagrams. What is needed are rules of a system that give us permission to perform certain manipulations. The validity of these rules presupposes the semantics of the system.

As I showed in detail in the second chapter, this is where our predecessors (including Peirce) stopped. They had a strong intuition about how Venn diagrams should be used. However, they were not able to justify their intuition, since they did not have a semantic analysis. They were satisfied with calling these diagrams the best *aids* to our monadic reasoning. Others who have used Venn diagrams after Peirce have followed these predecessors' evaluation of Venn diagrams.

I provided a semantic analysis of Venn diagrams, which gives a justification of the transformation rules of the diagrams. Accordingly, in this system, the traditional worry about the misapplication of diagrams disappears. Under the guidance of the justified syntax, we know which diagrams are permissible and which steps are permitted.

In two earlier studies on diagrammatic systems, Don Roberts, in *The Existential Graphs of Charles S. Peirce*, and John Sowa, in *Conceptual Structures*, independently proved that Peirce's system of existential graphs is sound and complete. Roberts proved these important results by way of first-order logic. He showed the existence of a translation between symbolic logic and the existential graphs and demonstrated that since first-order logic is sound and complete, so is Peirce's system. Sowa adopts a game-theoretic approach to prove the soundness and the completeness of Peirce's system. However, since Sowa also presented a translation between first-order logic and Peirce's system, the result reached by the game-theoretic approach is not surprising at all. These two works are very important in that they showed the soundness and the completeness of a diagrammatic system. In spite of this importance, little attention has been paid to these achievements. I think there are two main reasons for this unfair treatment. One is that Peirce's existential graphs have not been used much, mainly because they lost visual power when a diagrammatic device for negation was introduced. The other is that the correspondence between first-order logic and Peirce's system wrongly confirmed another prejudice against diagrams.

This further prejudice against nonlinguistic representation systems sees any nonlinguistic system as a translation or an illustration of a linguistic representation system. This belief leads one to regard a nonlinguistic representation system as, at best, a limited mirror of a linguistic representation system which is proved to be sound and complete. Accordingly, the rules of even a successful nonlinguistic system are taken to be subservient to the rules of a sound and complete linguistic system of which the former is an illustration. These beliefs have discouraged any attempt to explore representation systems other than linguistic ones.

Unfortunately, Roberts's and Sowa's works have incorrectly convinced some people that a system can be sound and complete only when it is a correct illustration of symbolic logic. This has led to the further mistaken belief that soundness and completeness are intrinsic to linguistic representation only. This is a prejudice for symbolic systems that holds that only linguistic systems are valid intrinsically and that other forms of systems can be valid only in a derivative sense.

Some might find my work on Venn diagrams parallel to that of Roberts or Sowa, claiming that my rules seem to be translations of well-known symbolic ones. They might think I transcribed one system to another with the result ending up being more complicated. It is true that some of my transformation rules are so similar to some inference rules of symbolic systems that the former almost seem to be copies of the latter. However, it is not true that these visual rules are copies of symbolic ones, even though they are very similar. The similarity arises not from copying but from the nature of our valid reasoning.

The subject of logic is reasoning itself. We may reason about many kinds of affairs – human beings, animals, pictures, life, or relations among these things, and so on – but logic is the study of valid reasoning, whatever we reason about. This is why logic is sometimes said to be subject-neutral. When our reasoning is about so many different kinds of things, how do we study the valid reasoning universal to these many different affairs? Before trying to answer this question, let us consider an example. We believe that the following reasoning is valid:

All bachelors are men. Tom is a bachelor.
Therefore, Tom is a man.

What makes this reasoning valid is *not* the relation among these three English sentences, *but* the relation among the pieces of information these sentences convey. The matter becomes clearer if we consider a first-order language:

$$\frac{\forall x(Bx \rightarrow Mx) \wedge Bt}{Mt}$$

Why is this reasoning valid? Some might say that this is valid since we can derive formula Mt from formula $\forall x(Bx \rightarrow Mx) \wedge Bt$. Of course, one has to add the observation that the derivation system is sound. This answer is not wrong. However, this answer might be considered to be circular, since soundness requires an understanding of validity. A more straightforward answer would be as follows: This reasoning is valid, since we may extract the information that formula Mt conveys from the information that formula $\forall x(Bx \rightarrow Mx) \wedge Bt$ conveys. Even if we did not know whether first-order logic is sound, this reasoning should still be considered valid. If we have a sound and complete system, we may figure out valid or invalid reasoning in a more mechanical way and, accordingly, in a more secure way, especially for complicated reasoning. This is one of the reasons why logicians are interested in sound and complete systems.

Whether the information in question is conveyed by means of English or by means of a first-order language does not matter to the essence of valid reasoning. To make this point clear, let us consider a rather radical example: Suppose we have a very inconvenient and not so useful language in which we have some logical constants with P, Q, and R among them. Suppose P means that every bachelor is a man, Q means that Tom is a bachelor, and R means that Tom is a man. Since these are logical constants, the meanings of these symbols are fixed. Now, if we infer R from P and Q, then we should consider this reasoning valid, *not* because we can translate this language into a more convenient first-order language, *but* because there is a necessary relation among the pieces of information these constants convey. If the essence of valid reasoning lies in the relation among pieces of information, we should be able to include any representation system in the scope of logic. As long as a system represents information, we should be able to judge whether or not that system is sound and complete. That representation system does not have to be linguistic. This is one of the main points I have tried to make in my work through the semantic analysis of a visual (nonlinguistic) representation system. This point, hopefully, will contribute to the following project which is now being undertaken by some logicians.

Emphasizing valid reasoning as a process of extracting information, not as a process of manipulating symbols, Barwise and Etchemendy turn our attention to heterogeneous reasoning. *Hyperproof* is their project in which they attempt to generalize our heterogeneous inference and to develop a mathematical system to deal with our multimodal reasoning.

In making inferences in ordinary life, human beings make use of information conveyed in many different kinds of forms, not just symbolic form. Barwise and Etchemendy correctly point out that the failure to realize the importance of multimodal reasoning is the basis for some complaints about traditional logic classes. One of the main criticisms against the traditional logic class is that learning logic in class does not help us perform valid reasoning despite the fact that logic is the study of valid reasoning. What we teach in logic class is to manipulate symbols, which is only one of many ways to handle information. The reasoning taught in logic class is single-mode reasoning, whereas ordinary reasoning is multimodal reasoning.

As noted before, the tradition in logic has been unfavorable to visual as opposed to linguistic representation systems. Venn diagrams are a particularly well-understood but previously unformalized system of visual representation. I have shown how this Venn system is sound and complete and thus hope to have contributed to clearing away one obstacle to the acceptance of logic as the study of multimodal reasoning.[1]

Even though this work is limited to two versions of Venn diagrams, now we know that nothing intrinsic about diagrams prevents us from using diagrams as valid proofs. I hope this will be an inspiring, albeit small, step that motivates others to work on other visual systems as well.

In the last chapter, I also tried to answer the question about the distinction between diagrammatic and linguistic systems. However, the discussion there is far from being a complete answer. A clear distinction between different kinds of representation systems, if there is any, is needed not just for theoretical purposes but also for practical uses. If we have a clear idea about the nature of linguistic and diagrammatic representation, then we might be able to choose a better system, depending upon the nature of the given information. Also, we might be able to discuss the following issues in a more substantial way: the limits and the advantages of diagrammatic representation, design principles for visual interaction, evaluation of visual representation, and so on. All these issues are beyond the scope of this work. However, I hope this work will contribute to research involving visual representation.

[1] Following up this work, a case study on heterogeneous inference has been carried out by Eric Hammer [13]. Hammer combines two different kinds of systems – Venn diagram systems which I developed and Fitch-style systems – to produce a heterogeneous system, which he also proved to be sound and complete.

Appendix

Theorem 1 *This extended function $\bar{s}$ has the following properties:*

(1) $\bar{s}(A) = s(A)$ *if* $A \in BRG$.
(2) $\bar{s}(A) = \bar{s}(A_1) - \bar{s}(A_2)$ *if* $A = A_1 - A_2$.
(3) $\bar{s}(A) = \bar{s}(A_1) \cap \bar{s}(A_2)$ *if* $A = A_1\text{-}and\text{-}A_2$.
(4) $\bar{s}(A) = \bar{s}(A_1) \cup \bar{s}(A_2)$ *if* $A = A_1 + A_2$.

Proof (Case 1) Suppose that A is a basic region. We show that $\bar{s}(A) = s(A)$.

(First we show $\bar{s}(A) \subseteq s(A)$.) Suppose that $a \in \bar{s}(A)$. Then, there is a minimal region, say, A_{mr}, such that $a \in s'(A_{mr})$, where A_{mr} is part of A. Since A_{mr} is a minimal region of A,

$$A_{mr} = (A\text{-}and\text{-}B_1^+\text{-}and\text{-}\cdots\text{-}and\text{-}B_i^+) - (B_1^- + \cdots + B_j^-), \quad (1)$$

where B_n^+ are the basic regions of which A_{mr} is a part and B_n^- are the basic regions of which A_{mr} is not a part.

By the definition of s',

$$s'(A_{mr}) = (s(A) \cap s(B_1^+) \cap \cdots \cap s(B_i^+)) - (s(B_1^-) \cup \cdots \cup s(B_j^-)). \quad (2)$$

By the assumption that $a \in s'(A_{mr})$ and equation (2),

$$a \in (s(A) \cap s(B_1^+) \cap \cdots \cap s(B_i^+)) - (s(B_1^-) \cup \cdots \cup s(B_j^-)). \quad (3)$$

Accordingly, $a \in (s(A) \cap s(B_1^+) \cap \cdots \cap s(B_i^+))$. Therefore, $a \in s(A)$.

(Now we show that $s(A) \subseteq \bar{s}(A)$.) Suppose that $a \in s(A)$. Let $B_1^+, \ldots, B_i^+$ be the basic regions (other than A) such that $a \in s(B_n^+)(1 \leq n \leq i)$. Let $B_1^-, \ldots, B_j^-$ be the basic regions such that $a \notin s(B_n^-)(1 \leq n \leq j)$. Accordingly,

$$a \in (s(A) \cap s(B_1^+) \cap \cdots \cap s(B_i^+)), \quad (4)$$

$$a \notin (s(B_1^-) \cup \cdots \cup s(B_j^-)). \tag{5}$$

By equations (4) and (5),

$$a \in (s(A) \cap s(B_1^+) \cap \cdots \cap s(B_i^+)) - (s(B_1^-) \cup \cdots \cup s(B_j^-)). \tag{6}$$

By equation (6), we know that

$$a \in s'(A_{mr}) \tag{7}$$

for minimal region A_{mr} such that

$$A_{mr} = (A\text{-}and\text{-}B_1^+\text{-}and\text{-}\cdots\text{-}and\text{-}B_i^+) - (B_1^- + \cdots + B_j^-).$$

By the definition of $\bar{s}$ and (7),

$$a \in \bar{s}(A). \tag{8}$$

(Case 2) Suppose that $A = A_1 + A_2$. We want to show that $\bar{s}(A) = \bar{s}(A_1) \cup \bar{s}(A_2)$. By the definition of $\bar{s}$,

$$\bar{s}(A) = \bigcup\{s'(A_{mr}) \mid A_{mr}\text{is a minimal region that is a part of } A\}, \tag{9}$$

$$\bar{s}(A_1) = \bigcup\{s'(A_{mr}) \mid A_{mr}\text{is a minimal region that is a part of } A_1\}, \tag{10}$$

$$\bar{s}(A_2) = \bigcup\{s'(A_{mr}) \mid A_{mr}\text{is a minimal region which is a part of } A_2\}. \tag{11}$$

Since $A = A_1 + A_2$, every minimal region of A is a minimal region of either A_1 or A_2, *and* any minimal region which is either part of A_1 or A_2 is a minimal region of A. Therefore, by equations (9), (10) and (11), $\bar{s}(A) = \bar{s}(A_1) \cup \bar{s}(A_2)$.

(Case 3) and (Case 4): They are similar to case 2. □

Theorem 4 *For any regions R_1 and R_2 of a wfd, if R_1 and R_2 do not overlap each other, then the set represented by R_1 and the set represented by R_2 are disjoint.*

Proof Let R_1 and R_2 be regions of a *wfd D* such that R_1 and R_2 do not overlap. Let U be the region enclosed by the rectangle of *wfd D*. Since R_1 and R_2 do not overlap, R_2 must be a part of $U - R_1$. By Theorem 3,

$$\bar{s}(R_2) \subseteq \bar{s}(U - R_1). \tag{1}$$

I Claim

$$\bar{s}(R_1) \cap \bar{s}(U - R_1) = \emptyset.$$

This follows from

$$\overline{s}(R_1) \cap \overline{s}(U - R_1) = \overline{s}(R_1) \cap (\overline{s}(U) - \overline{s}(R_1)) = \emptyset.$$

Suppose that $\overline{s}(R_1) \cap \overline{s}(R_2) \neq \emptyset$. Then, there is an element x such that $x \in \overline{s}(R_1)$ and $x \in \overline{s}(R_2)$. By (1), $x \in \overline{s}(U - R_1)$. Therefore, $x \in \overline{s}(R_1)$ and $x \in \overline{s}(U - R_1)$. Hence, $\overline{s}(R_1) \cap \overline{s}(U - R_1) \neq \emptyset$, which contradicts the subclaim. Therefore, $\overline{s}(R_1) \cap \overline{s}(R_2) = \emptyset$. □

Theorem 5 *For every region R of a wfd, the set represented by R is the union of the sets represented by the minimal regions of which R consists. (I.e.,* $\overline{s}(R) = \bigcup\{\overline{s}(R_{mr}) \mid R_{mr}$ *is a minimal region and* R_{mr} *is a part of* $R\}$.*)*

Proof It follows by induction on the number of minimal regions of which R consists

(Basis Case) Let R consist of only one minimal region, R_{mr}. Hence, $R = R_{mr}$. Therefore, $\overline{s}(R) = \bigcup\{\overline{s}(R_{mr})\}$.

(Inductive Step) Inductive Hypothesis: If R consists of n minimal regions, then $\overline{s}(R) = \bigcup\{\overline{s}(R_{mr}) \mid R_{mr}$ is a minimal region and R_{mr} is a part of $R\}$.

Suppose that R consists of $n+1$ minimal regions. Let $R_1, \ldots, R_n, R_{n+1}$ be the minimal regions. Hence, $R = R_1 + \cdots + R_n + R_{n+1}$. Let $R' = R_1 + \cdots + R_n$. So, $R = R' + R_{n+1}$. By definition of $\overline{s}$,

$$\overline{s}(R) = \overline{s}(R') \cup \overline{s}(R_{n+1}). \tag{1}$$

Since R' consists of n minimal regions, by the inductive hypothesis

$$\overline{s}(R') = \overline{s}(R_1) \cup \cdots \cup \overline{s}(R_n). \tag{2}$$

By equations (1) and (2),

$$\overline{s}(R) = \overline{s}(R_1) \cup \cdots \cup \overline{s}(R_n) \cup \overline{s}(R_{n+1}). \tag{3}$$

Therefore, $\overline{s}(R) = \bigcup \{\overline{s}(R_{mr}) \mid R_{mr}$ is a minimal region and R_{mr} is a part of $R\}$. □

Theorem 10 *If* $RF(D') \subseteq_{cp} RF(D)$, *then* $\{D\} \vdash D'$.

Proof If $BR(D') \subseteq_{cp} BR(D)$, then by Theorem 7 and Theorem 9 (in order), $\{D\} \vdash D'$. If it is not the case that $BR(D') \subseteq_{cp} BR(D)$, then apply the rule of introduction of a basic region to D to get D'' such that $BR(D') \subseteq_{cp} BR(D'')$. Then, use Theorem 7 and Theorem 9. □

References

[1] Barwise, Jon. Heterogeneous Reasoning. In *Conceptual Graphs and Knowledge Representation*, eds. G. Mineau, B. Moulin, and J. Sowa. New York: Springer Verlag, in press.

[2] Barwise, Jon, and John Etchemendy. Visual Information and Valid Reasoning. In [29], pp. 9–24.

[3] Barwise, Jon, and John Etchemendy. Information, Infons and Inference. In *Situation Theory and Its Application I*, vol. 1, eds. Cooper, Mukai, and Perry, pp. 33–78. Stanford: CSLI, 1990.

[4] Barwise, Jon, and John Etchemendy. Hyperproof: Logical Reasoning with Diagrams. In *Reasoning with Diagrammatic Representations*, ed. N. Hari Narayanan. Menlo Park: AAAI Press (in press).

[5] Barwise, Jon, and John Perry. *Situations and Attitudes*. Cambridge: MIT Press, 1983.

[6] Brumfiel, C., R. Eicholz, and M. Shanks. *Geometry*. Reading, MA: Addison-Wesley, 1960.

[7] Dori, Dov. Dimensioning Analysis: Toward Automatic Understanding of Engineering Drawings. *Communications of the ACM* 35, 10 (1992) pp: 92–103.

[8] Eisenberg, Theodore and Tommy Dreyfus. On the Reluctance to Visualize in Mathematics. Unpublished manuscript available from Weizmann Institute of Science, Rehovot, Israel, 1990.

[9] Euler, Leonhard. *Briefe an eine deutsche Prinzessin*. Braunschweig: Vieweg, 1986.

[10] Gardner, Martin. *Logic Machines and Diagrams*. 2d ed. Chicago: University of Chicago Press, 1982.

[11] Goodman, Nelson. *Languages of Art*. 2d ed. Indianapolis: Hackett, 1976.

[12] Hadamard, J. *The Psychology of Invention in the Mathematical Field*. New York: Dover, 1945.

[13] Hammer, Eric. Reasoning with Sentences and Diagrams. *Notre Dame Journal of Formal Logic*, (in press).

[14] Kneale, William, and Martha Kneale. *The Development of Logic*. Oxford: Clarendon Press, 1986.

[15] Lambert, Johann Heinrich. *Neues Organon* I. Berlin: Akademie Verlag, 1990.

[16] Larkin, J. and H. Simon. Why a Diagram Is (Sometimes) Worth Ten Thousand Words. *Cognitive Science* 11 (1987): 65–99.

[17] Ogilvy, C. Stanley. *Excursions in Geometry*. New York: Oxford University Press, 1969.

[18] Peirce, Charles Sanders. *Collected Papers*, volume 4. Ed. Charles Hartshorne and Paul Weiss. Cambridge, MA: Harvard University Press, 1933.

[19] Polythress, V., and H. Sun. A Method to Construct Convex Connected Venn Diagrams for Any Finite Numbers of Sets. *Pentagon*. 31 (Spring 1972), pp: 80–3.

[20] Quine, Willard Van Orman. *Methods of Logic*. New York: Holt, Rinehart & Winston, 1964.

[21] Roberts, Don. *The Existential Graphs of Charles S. Peirce*. The Hague: Mouton, 1973.

[22] Salmon, Merrilee. *Introduction to Logic and Critical Thinking*. 2d ed. San Diego: Harcourt Brace Jovanovich, 1989.

[23] Shin, Sun-Joo. An Information-Theoretic Account of Valid Reasoning with Venn Diagrams. In *Situation Theory and Its Applications*, vol. 2, eds. Barwise, Gawron, Plotkin, and Tutiya, pp. 581–605. Stanford: CSLI, 1992.

[24] Shin, Sun-Joo. Peirce and the Logical Status of Diagrams. *History and Philosophy of Logic*, 15 (1994): 45–68.

[25] Sowa, John. *Conceptual Structure: Information Processing in Mind and Machine*. Reading, MA: Addison-Wesley, 1984.

[26] Tennant, Neil. The Withering Away of Formal Semantics? in *Mind and Language* 1, 4. (1986): 302–18.

[27] Venn, John. *Symbolic Logic*. 2d ed. New York: Burt Franklin, 1971.

[28] Wittgenstein, Ludwig. *Tractatus Logico-Philosophicus*. Trans. D. F. Pears and B. F. McGuiness. London: Routledge & Kegan Paul, 1961.

[29] Zimmerman, Walter, and Steve Cunningham, eds. *Visualization in Teaching and Learning Mathematics*. Washington, DC: Mathematical Association of America, 1990.

Index